T0300252

Inland Waterway Transportation

Inland Waterway Transportation explores how tools of economic analysis can improve the efficiency of both public and private investment in inland waterway transportation. Originally published in 1969, this study investigates how waterway transportation has been affected by public operating policy, costs and charges for the use of waterways in the United States as well as the impact of relationships central to waterway policy and individual firms such as the effect of the waterway environment on a firm's efficiency. This title will be of interest to students of Environmental Studies and professionals.

Inland Waterway Transportation

Studies in Public and Private Management and Investment Decisions

Charles W. Howe, Joseph L. Carroll, Arthur P. Hurter, Jr., William J. Leininger, Steven G. Ramsey, Nancy L. Schwartz, Eugene Silberberg and Robert M. Steinberg

RFF PRESS
RESOURCES FOR THE FUTURE

First published in 1969
by Resources for the Future, Inc.

This edition first published in 2016 by Routledge
2 Park Square, Milton Park, Abingdon, Oxon, OX14 4RN
and by Routledge
711 Third Avenue, New York, NY 10017

Routledge is an imprint of the Taylor & Francis Group, an informa business

Publisher's Note
The publisher has gone to great lengths to ensure the quality of this reprint but points out that some imperfections in the original copies may be apparent.

Disclaimer
The publisher has made every effort to trace copyright holders and welcomes correspondence from those they have been unable to contact.

A Library of Congress record exists under LC control number: 71085340

ISBN 13: 978-1-138-95496-0 (hbk)
ISBN 13: 978-1-315-66649-5 (ebk)

INLAND WATERWAY TRANSPORTATION

STUDIES IN PUBLIC AND PRIVATE

MANAGEMENT AND INVESTMENT DECISIONS

Photo courtesy of *Steelways* (March/April 1968)

INLAND WATERWAY TRANSPORTATION

Studies in Public and Private Management and Investment Decisions

by

Charles W. Howe
Joseph L. Carroll
Arthur P. Hurter, Jr.
William J. Leininger

Steven G. Ramsey
Nancy L. Schwartz
Eugene Silberberg
Robert M. Steinberg

RESOURCES FOR THE FUTURE, INC.

Distributed by The Johns Hopkins Press

Baltimore and London

Resources for the Future, Inc.
1755 Massachusetts Avenue, N.W., Washington, D.C. 20036

Resources for the Future is a non-profit corporation for research and education in the development, conservation, and use of natural resources. It was established in 1952 with the co-operation of the Ford Foundation and its activities since then have been financed by grants from the Foundation. Part of the work of Resources for the Future is carried out by its resident staff, part supported by grants to universities and other non-profit organizations. Unless otherwise stated, interpretations and conclusions in RFF publications are those of the authors; the organization takes responsibility for the selection of significant subjects for study, the competence of the researchers, and their freedom of inquiry.

The authors are associated with the following organizations: Charles W. Howe—Director, Water Resources Program, Resources for the Future, Inc.; William J. Leininger—Operations Research, Inc.; Arthur P. Hurter, Jr.—Professor of Management Science, Northwestern University; Nancy L. Schwartz—Associate Professor of Economics, Carnegie-Mellon University; Robert M. Steinberg—formerly Research Associate, Resources for the Future, Inc., now Chief, Statistical Methodology and Procedures Section, Division of Research and Statistics, Federal Reserve Board of Governors; Joseph L. Carroll—Head, Transportation Research Division, Pennsylvania Transportation and Traffic Safety Center, The Pennsylvania State University; Steven G. Ramsey—Lieutenant, U.S. Army; Eugene Silberberg—Assistant Professor of Economics, University of Washington.

The figures were drawn by Clare and Frank J. Ford.

RFF staff editors: Henry Jarrett, Vera W. Dodds, Nora E. Roots, Sheila M. Ekers.

PREFACE

This volume is the outcome of a program of studies designed to develop tools of economic analysis that could be used to improve the economic efficiency of public and private investment in shallow-water inland transportation. As in any transport system in which the right-of-way is publicly provided for the use of privately owned carriers, there is an obvious and close interdependence between the volume and form of resources invested by public authority and the operating conditions (and consequently costs) faced by the private carriers. Thus, the bargeline firm is presented with quite a different set of technical options regarding equipment types and tow size by deep, wide, straight waterways, with large chambered locks, than by less improved waterways.

Operating conditions faced by carriers in the industry are not only affected by public investments in waterway improvements; public operating policies also have important immediate effects. The rules governing locking priorities, for example, in large part dictate the maximum sizes of tows on a particular waterway, for tows that require more than a "double locking" (being broken and passed through the lock in two sections) must let others pass through the locks first. The present rules which specify that pleasure craft shall not have to wait for more than three commercial lockages also can affect the costs of commercial operations.

Operating policies relating to the multiple purpose features of river systems have an important impact on the private carrier. If pools are kept low to provide flood protection capacity, the draft of vessels may be restricted. If hydroelectric power (especially peaking power) is generated from stored waters, the surges can interfere seriously with navigation and restrict the draft of vessels to the minimum depth occurring during the periods of low generation while water is being stored.

There can also be direct conflict between the operation of the system for irrigation and its navigation uses, since both require releases from storage during the season of low flow. Such a situation is discussed at length in Appendix B to Chapter 1.

Finally, the bargeline firm is affected by the conditions of access to navigation on the waterway and, in particular, whether or not charges are made for the use of the waterway. This important public operating policy affects the firm both financially and through its effect on the degree of congestion. Present policies of free access result in excessive congestion. Carriers of cargoes for which waterway transport holds a differential advantage greater than other modes of transport lose more from added congestion than they would from paying an optimum toll. This issue is analyzed in Chapter 5.

It is clear from these interactions that effective public decision making with regard to investment and operating policies must take into account the reactions of private carriers and that, similarly, effective private decisions must be based on present and forecast public policies.

The studies which follow deal with shallow-water transportation as carried out through contemporary barge and towboat technology on the rivers, canals, and intracoastal waterways of the United States. Transportation on the Great Lakes and coastal deep-draft shipping are not included. Magnitudes and relationships central to public waterway policy and of relevance to the individual firm are investigated: i.e., the technology of modern barge transport and its relationship to publicly provided rights-of-way; the impact of the waterway environment on the efficiency of the firm; the relationship of private equipment investment by the firm to the transport demand pattern; the measurement of benefits from waterway improvement, including the prediction and structure of congestion costs; and the prediction of route-by-route demands for transport.

The studies originated in several places. In 1962, a program of research was initiated at Purdue University by Charles W. Howe, William J. Leininger, Eugene Silberberg, and Nancy L. Schwartz. In the same year a parallel study, supported by a grant from Resources for the Future, Inc., was started at the Transportation Center at Northwestern University. Arthur P. Hurter, Jr. carried out his work at the Center. In 1965, it was decided that Howe's work on the production function and Joseph L. Carroll's studies on locking delay simulation could usefully be brought together to produce a computer simulation program capable of detailed simulation of navigation operations on a major segment of a waterway system. This work was undertaken by Carroll and Steven G. Ramsey at Pennsylvania State University and by Howe and Robert M. Steinberg at Resources for the Future. All these studies have been substantially revised before being included in this volume.

Many persons have assisted the authors greatly by providing counsel and data. Without their help, none of the present studies could have been completed. While their assistance was indispensable, those whose help is acknowledged below are to be exonerated completely from the shortcomings of the present volume. Further, it should be clearly noted that this acknowledgement of indebtedness in no way implies agreement of the individual (or of his institution) with any conclusions or proposals set forth herein.

The late Captain A. C. Ingersoll, Jr., former President of Federal Barge Lines, was a leader of the bargeline industry for many years and an inspiration to all who knew him. His enthusiasm for his industry and straightforward analysis of its problems set standards by which all shall be measured.

Many other persons associated with the bargeline industry lent their help at crucial times: Noble Parsonage and E. C. Ross of Federal Bargelines; J. W. Hershey, Floyd Blaske, W. Armin Willig, and Charles E. Peters of American Commercial (Barge) Lines; Alex S. Chamberlain of Ashland Oil Company; Louis R. Fiore and David T. Sheehy of the Ohio River Company; Floyd A. Mechling and John W. Oehler of A. L. Mechling Barge Lines; C. J. Bogman of the American Bridge Division, U.S. Steel Corporation; and A. M. Martinson, Jr. and I. C. Douthwaite of the Dravo Corporation.

Among academic colleagues who were of great help were Leon N. Moses of Northwestern University; Michael F. Brewer, Allen V. Kneese, and John V. Krutilla of RFF; and Marianne Yates and Mary Roy, Transportation Center Librarians. Invaluable aid in data gathering and initial data processing was rendered by Arie Beenhakker, Paul De Schutter, and Dean C. Smith—at that time graduate students in economics at Purdue University.

The U.S. Army Corps of Engineers, particularly through the Office of the Chief of Engineers, the Board of Engineers for Rivers and Harbors, and the Ohio River Division, provided needed advice and data: thanks are due Nathaniel A. Back, William J. Rhodes, Edmund H. Lang, Leonard T. Crook, Eric E. Bottoms, T. P. Bailey (retired), Alan R. Chandler, W. R. McClintock, and E. E. Abbott.

A special debt is owed to Braxton B. Carr and The American Waterways Operators, Inc. for permission to use the excellent descriptive materials on the bargeline industry contained in Appendix A of Chapter 1. Another specific debt is owed to T. Waara, Missouri River Division, U.S. Army Corps of Engineers, for the materials on multiple purpose river management in Appendix B of Chapter 1.

The authors wish to express their thanks to Sheila Ekers, who very creatively edited the manuscript, and to Dee Stell, who efficiently and patiently typed the many different versions that preceded the final product.

The authorship of the several chapters is as follows: Howe wrote Chapter 1 (except for the Appendixes) and derived the technological production and cost functions of Chapter 2. Leininger derived the empirical production functions of Chapter 2. Howe carried out the returns-to-scale studies of Chapter 3 while Hurter performed the cost analyses. Chapter 4 represents the research of Schwartz. Attributions are most difficult for Chapter 5. The general outline of the model was Howe's, but Carroll helped in the early formulation. The program writing was started by Ramsey under Carroll's supervision. Steinberg then produced the heart of the model through his excellent programming, criticism, and ideas. The demand model of Chapter 6 was constructed and estimated by Silberberg.

After the book had gone to press, it was learned that Steven Ramsay had been killed in Vietnam. His coauthors wish to dedicate the book to him and to the promise that was lost.

CONTENTS

LIST OF TABLES

LIST OF FIGURES

INLAND WATERWAY TRANSPORTATION

STUDIES IN PUBLIC AND PRIVATE

MANAGEMENT AND INVESTMENT DECISIONS

1

INTRODUCTION AND SUMMARY: A FRAMEWORK FOR PUBLIC DEVELOPMENT AND MANAGEMENT OF THE INLAND WATERWAYS

Although barges carry about 9.5 per cent of all ton-miles of freight in the United States[1] and expenditures on the federal rivers and harbors program amounted in 1967 to over one billion dollars—ranking only behind defense, agriculture, and space in federal government outlays—inland waterway transport and the programs of public investment that have supported the industry's remarkable growth since World War II have received surprisingly little attention.

This volume provides certain tools that can help to improve the efficiency of decisions on the allocation of public resources to navigation and of private resources to towboats and barges. Indirectly, improvement in both design and management procedures for navigation projects and navigation features of multiple purpose projects will assist in the formulation of better decisions on the allocation of public resources among the major fields of government expenditure.

The multiple purpose aspects of river development, including navigation, power generation, flood control, irrigation, recreation, fish and wildlife, and general aesthetic features, have been treated in some detail from the public investment point of view. Eckstein [1961] and Krutilla and Eckstein [1958] have developed the analytical framework for project evaluation and the design of multiple purpose systems. Much of this basic analytical work has been translated into guidelines for practice in the "Greenbook"[2] and in Senate Document 97.[3] Much more detailed consideration of the design of water resource (river) systems has been presented by Maass et al. [1962]. Meyer et al. [1959] have analyzed waterway transportation from the point of view of a co-ordinated national transport system and a rational allocation of traffic among the various modes.

However, there have been neither detailed analytical methods for determining optimum waterway investment and management procedures nor satisfactory procedures for the analysis of the demand for the use of particular waterways. As a result, questionable estimates of future traffic on existing or projected waterways have been incorporated into public waterway investment evaluations. New waterways and those being improved are still largely designed on a project-by-project basis. This is partly because there has been no framework for the

[1] See Table 3, p. 18.
[2] *Proposed Practices for Economic Analysis of River Basin Projects*, report prepared by the Subcommittee on Evaluation Standards of the Inter-Agency Committee on Water Resources (Washington: Government Printing Office, 1958).
[3] *Policies, Standards, and Procedures in the Formulation, Evaluation, and Review of Plans for Use and Development of Water and Related Land Resources*, S. Doc. 97, 87 Cong. 2 sess. (1962).

analysis of alternative designs of larger systems. The management of existing waterways in terms of traffic regulation and operating rules for locks, while not obviously inefficient, has rested on a traditional policy of free access and service to all users, both commercial and recreational, and certain rules of thumb concerning lock operation. As the volume of traffic on the waterways increases and as the costs of physical expansion of the channels and locks of the inland waterway system rise, it becomes increasingly important that the waterways should be designed and managed as a system. The studies of this volume take steps in this direction.

THE ECONOMIC EFFICIENCY MODEL
FOR WATERWAY INVESTMENT AND MANAGEMENT

Resources are allocated in an economically efficient way when the pattern of their use maximizes the present value of the national income, broadly construed. Such an allocation must result from a sequence of public and private decisions at different levels: decisions on the division of resource use between the private and public sectors, among the major programs within the public sector (e.g., education, defense, health, water development); among alternative systems within each public program (e.g., among river basins); among the components or outputs of each system (navigation, power, water supply, flood control, water quality management); and among alternative technological features for producing each output (e.g., flood water impoundment versus floodproofing of buildings).

Most water resource development has been left to public authority for two reasons: the "public good" nature of many water system outputs and the pervasiveness of externalities. "Public goods" are those whose use or enjoyment by one party in no way diminishes the possibility of enjoyment by others: the aesthetic enhancement of a river valley is an example. Externalities are those effects, like the downstream impacts of pollution, that would not be taken into the private utility or profit calculations of the firm or local authority.[4] While the private sector relies on competitive pressures to enforce efficient decisions, there is no competition for the typical activities of public authorities. Since no market exists for the public goods, explicit analyses of benefits and costs are necessary.

The measurement of benefits from water development projects is based conceptually upon *demand curves* representing schedules of incremental values to the users of different quantities of each output from the project. To be relevant to the planning of navigation facilities, such demand curves must be defined for particular waterways or waterway segments and not at higher levels of aggregation.

Transportation of materials and products occurs because differences exist in production opportunities and demand structures among regions. The demands

---

[4] These concepts are more fully explored in Krutilla and Eckstein [1958], Eckstein [1961], and Kneese [1964].

for use of the waterways are not fully represented by the tonnage flows among regions; they must be expressed in terms of the numbers of units (barge tows) of different sizes that will use the particular waterway under different conditions of operating costs, prices charged for use of the waterway, and congestion. The transformation of gross commodity flows into movements of tows of various sizes following specific itineraries is performed by the bargeline firms as they schedule their boats and barges to execute the commodity movements demanded by their customers. This transformation is dependent upon: (1) the gross commodity flows and their origin-destination pattern; (2) the number of bargeline firms and the locations of their particular customers; (3) the available transport technology and the conditions of the rivers and canals; and (4) the scheduling procedures used by the bargeline firms. Naturally, there will be a feedback from the operating costs of the bargelines through the rates they charge shippers to the volume of gross commodity flows, so that, over the longer term, all of these things are determined simultaneously.

Once the demand for the use of a waterway is specified as a function of operating conditions (including user charges, if any), the analysis of optimum design and management requires a knowledge of the public and private costs that are incurred in developing and using the waterway. Four types of costs are incurred during or prior to actual travel on the waterway: (1) public capital and operating-and-maintenance (O&M) costs of constructing and operating the waterway; (2) opportunity costs in terms of forgone alternative outputs of the multiple purpose waterway system (e.g., use of the water for power or irrigation); (3) private "straight-through" operating costs which would be incurred by tows if no congestion were present (i.e., if all channels and locks were always ready to receive a tow upon arrival); (4) congestion costs. This partitioning of costs permits the demand function to be defined in terms of the tow's willingness to pay for passage along the waterway net of straight-through operating costs, rather than being dependent upon the particular volume of traffic. Congestion costs are thus kept separate as a function of traffic volume. This leads to a logically deterministic model of the demand-supply relationship for the waterway (discussed in greater detail in Chapter 5). The problem now is to bring together these elements in a model of the waterway system in order to determine the optimum amount of public investment in the physical structures of the system and the optimum volume of traffic.

Let us assume that the demand function is known as a function of both the average number of tows using the waterway per day and the extent of public investment in waterway improvements. The demand function indicates the willingness to pay for use of the waterway, and the area under the demand function can be interpreted as the *gross* benefits forthcoming from using a waterway of the specified characteristics at the indicated rate of traffic flow. The problem of optimum waterway design and management, then, is: How can net benefits be maximized by selecting the amount to be invested in waterway improvements and the volume of traffic to use the waterway? Net benefits are defined as the difference between the aggregate willingness to pay as measured by the demand

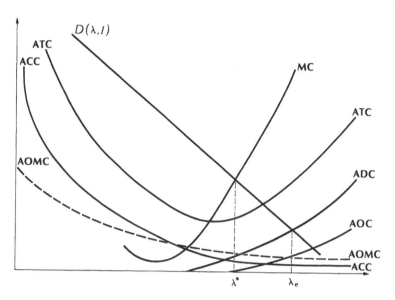

FIGURE 1.   Demand for and costs of inland waterway transportation.

$D\,(\lambda,\,I)$ = Demand function.
ATC = Average total cost curve.
ACC = Average capital cost curve.
$AOMC$ = Average operating and maintenance cost curve.
MC = Marginal cost curve.
ADC = Average delay (congestion) cost curve.
AOC = Average opportunity cost of water curve.

function and the sum of the following costs: the average annual capital costs; the annual operating and maintenance (O&M) costs; the value of other water system outputs forgone because of navigation; and congestion costs.[5] The essential features of the problem are exhibited in Figure 1, where the demand curve shows tow operators' willingness to pay for the use of the waterway and the cost curves represent the relevant costs that are being incurred either by

[5] Mathematically, this can be stated as

$$\underset{\lambda,\,I}{\text{maximize}}\left[\varphi(\lambda,\,I)\,=\,\int_0^{\lambda}D(\tilde{\lambda},\,I)d\tilde{\lambda}\,-\,K(I)\,-\,f(\lambda,\,I)\,-\,g(\lambda)\,-\,h(\lambda,\,I)\right],$$

where $D(\lambda,\,I)$ represents the willingness of tow operators to pay for use of the waterway as a function of the average number of tows λ and public waterway investment I; $K(I)$ is the annual equivalent capital cost; $f(\lambda,\,I)$ are the O&M costs; $g(\lambda)$, the opportunity costs of the water used; and $h(\lambda,\,I)$, the congestion costs.

It will be recalled that the marginal willingness to pay has been defined to be net of straight-through operating costs. The presence of I as an argument of the demand function presupposes an ordering of particular projects which would be executed when particular amounts of capital, I, were available for investment.

For a more detailed discussion of the use of average rates of demand when traffic flows exhibit large random variations, see Chapter 5. Here it is assumed that all tows can be considered homogeneous in terms of their impact on system performance.

$$K(I)\,=\,\frac{r\,I}{1\,-\,(1\,+\,r)^{-T}}\text{ when }r\text{ is the discount rate and }T\text{ is the life of the project.}$$

private operators (congestion costs); by the public waterway authority (capital and O&M costs); or by other parties, public and private, who are forced to forgo benefits because of the navigational use of water (e.g., forgone power output, irrigation, or water quality releases). All the costs are shown as averages; i.e., dollars per tow using the waterway.

The policy variables whose values are to be determined are I, the amount to be invested in waterway improvements, and λ, the rate of traffic flow expressed in tows per day. To take I as this policy variable naturally assumes a large amount of prior optimizing so that each volume of expenditure corresponds to the selection of whichever projects can best be undertaken for that cost. Taking λ as a policy variable assumes a type of control of traffic which has never been tried in this country—the control of traffic volume through some system of regulation. The system of traffic regulation most in keeping with the free market system is the imposition of a user charge.

A solution to the problem of selecting a socially optimal level of investment and traffic can perhaps best be described in two steps. First, assume that the physical waterway system is given (I is fixed). Then the optimum volume of traffic must be determined. Referring to Figure 1, there are good reasons to expect that uncontrolled access to the waterway will result in an average volume of traffic of λ_e because at this level, in the absence of a user charge, only the congestion costs are borne by the tow operators. Furthermore, every tow suffers in about the same way from congestion; i.e., each experiences congestion costs in an amount approximately equal to the average. Thus, bargelines will continue to put tows on the river until their marginal willingness to pay for the use of the waterway equals the average congestion costs experienced.

From the point of view of economic efficiency, however, the best volume of traffic to enforce would be λ^*. At this level, marginal (incremental) costs, both publicly and privately borne, equal the marginal willingness to pay for use of the waterway. Thus, λ^* is selected as a socially desirable level of traffic because, at any lesser volume of traffic, incremental willingness to pay exceeds the incremental costs imposed on society by another unit of traffic, implying that additional traffic would yield net benefits to society. At any greater volume of traffic than λ^*, the incremental social costs outweigh the incremental advantage to the bargelines, and some traffic should be discouraged.

The second step of the optimization requires the selection of an optimal investment program for the waterway; e.g., improvements in locks and channels, elimination of curves, elevation of bridges, etc. This decision cannot easily be represented graphically, but it represents the familiar exercise of a marginal benefit-cost analysis to determine how far the program of improvement should be carried. Since there are many ways in which any waterway could be improved physically, this task is in practice formidable.[6] The problem is further complicated by the shifting of the demand function over time. The problem is one of dynamic optimization in the face of rather "lumpy" investment oppor-

[6] The computer simulation model described in Chapter 5 is designed to permit the design team to experiment with different combinations of improvements.

tunities. The condition for optimization is that, for the last increment to the system, the present value of savings in straight-through operating costs plus savings in congestion costs should equal or just exceed the present value of the increase in O&M costs and capital costs.

SUMMARY AND GUIDE TO THE STUDIES

A description and brief summary of the studies to follow will be useful at this point. Appendix A to this chapter provides the reader who is not familiar with the inland waterway transportation industry with a brief discussion of the history, technology, and geographical pattern of the industry. Appendix B discusses the opportunity cost of water used for navigation in the Missouri Basin.

Chapter 2 is devoted to the technology of modern inland waterway transportation. A technologically based production function is derived to relate the performance of the tow to the characteristics of the (publicly provided) channel in which it operates. This analysis was supplemented by the empirical production functions of Leininger, based on log-book operating data.

It was found that the favorable effects of increased channel width on tow performance are largely exhausted at widths twice that of the tow, and that the favorable effects of increased depth are largely exhausted at depths four times that of the average flotilla draft. The use of the device known as the Kort nozzle was shown to result in an average increase of 12 per cent in a tow's rate of output (in cargo ton-miles per hour). The direction of travel was, quite naturally, found to have a much more dramatic effect on the rate of output of the tow during high water or "open river" conditions than when the waterway was in "pool stage." Leininger's analysis of Ohio River log-book data showed that the average upstream rate of output was only 47 per cent of the downstream rate during open river conditions, but that it was 86 per cent of the downstream rate when the river was in pool stage.

The production function and related cost analyses have shown that tows become increasingly efficient as their size increases (assuming proportional increases in the barge and boat-horsepower inputs) up to a critical size. Beyond this the flotilla resistance increases disproportionately because of the growth of the cross-sectional area of the tow relative to that of the channel. This critical size of tow is a function of the depth and width of the waterway. Thus, on the Lower Mississippi, increasing efficiency of tow operations probably extends beyond the maximum tow sizes currently used (e.g., 9,000 horsepower and up to forty barges), but on smaller waterways, the onset of decreasing efficiency is probably a very real operating constraint.

Chapter 3 investigates the extent to which the increasing efficiencies of larger tow size result in more efficient bargeline operations as the size of the firm grows. Sample evidence indicates that bargeline firms generally become more efficient as they grow, but that the degree of increasing efficiency (degree of increasing returns to scale) depends upon the type of waterway in which the firm's tows operate. Cost analysis also indicates an increase in the efficiency of the firm as its size increases.

Chapter 4 demonstrates a methodology for determining the size of the boat and barge fleets needed to carry out a particular pattern of cargo transport. The formulation of the problem incorporates point-to-point cargo demands which have probabilistic variations built into them, but the analysis is carried out in terms of the average (expected) cargo loads to be carried. The solutions of the derived mathematical model are then checked against a computer simulation of the problem that again actively incorporates the random variations in cargo loads. The proposed solutions appear to work and indicate that the method is capable of extension to real-world equipment problems.

Chapter 5 incorporates the technology of shallow-water transportation in a computer simulation model of navigation on a waterway system. The characteristics of the model, including the size of the system simulated, are variable and adaptable to realistic systems of up to ten sets of dual locks. Adaptation of the program to a computer with greater core storage would permit substantial expansion of the system simulated. The model permits a systems analysis of the impact of alternative physical and operating rule improvements, and also permits a detailed study of congestion in the system as a function of both the volume and characteristics of the traffic. The importance of the systems approach, as opposed to a project-by-project approach, is that the impacts of an improvement may be felt throughout the system or at points far removed from the actual improvement. It is demonstrated, for example, that at high traffic densities—when delay time and waiting lines at locks are long—the benefits accruing to the total system from alleviating the congestion at a particular lock (through construction of a larger lock or use of more efficient operating rules) are substantially less than the benefits measured only at the point of improvement. The congestion is, to some extent, transferred to other parts of the system.

It is demonstrated for hypothetical cases similar to the modernized reaches of the Ohio River that congestion costs can be substantial and that different congestion costs are imposed on the system by tows of different sizes. Marginal congestion costs differ substantially from average congestion costs at moderate traffic levels, indicating the desirability of some type of control over traffic volume. Procedures are described for determining approximately optimal user charges. The question of how large a system must be included in the model to incorporate all significant impacts of system changes remains largely unanswered, although the results of some experiments designed to generate information on this point are described.

Chapter 6 addresses the problem of predicting point-to-point commodity shipments for a waterway system. The problem is particularly difficult because of the large number of factors that must be taken into account. Lack of data precludes the estimation of individual "demand for barge transport" equations for all relevant origin-destination pairs. In the approach taken here, the U.S. waterway system is reduced to twelve regions and an attempt is made to predict the region-by-region tonnage movements through the use of a combination of econometric forecasting equations and a linear programming model for transport cost minimization. The potential benefits of this type of model stem from its ability to trace the impact on the entire barge transport system of changes in barge or rail rates and in regional economic activity. The evidence indicates that the model has a surprisingly good predictive ability and that further developmental work on models of this type is warranted.

APPENDIXES TO CHAPTER 1

A. INLAND WATERWAY TRANSPORTATION IN THE UNITED STATES

The following brief materials have been compiled from published sources in order to provide the reader with an understanding of the technology and operations of barge transport and the related public system of rivers and canals.

Historical Background[7]

From the days of the earliest settlements along the Atlantic seaboard, waterway transportation has played a major role in the nation's development. In retrospect three fairly distinct periods may be discerned.

The first of these began in the early seventeenth century with the use of the rivers in their natural condition as routes of exploration. Soon these rivers became the highways by which settlers gained access to the hinterlands. With settlement came production and the need to transport raw materials to the cities and manufactured goods to the frontier. This gave rise to efforts to improve the large rivers as highways of commerce, as well as to subsequent attempts to link them by man-made canals. The first major project—the Erie Canal, built by the state of New York—proved so profitable that it encouraged the states and private enterprise to construct a remarkable network of canals. The age of canal building overlapped the earlier years of the steamboat period, during which river transportation, particularly on the Mississippi and Ohio rivers, reached a stage of development not to be equaled until the third, or modern, era. Between 1811 and the Civil War, the steamboat, or packet, revolutionized river transportation. In 1860, an average of ten steamboats arrived at New Orleans each day, and during that year delivered cargoes valued at half a billion of today's dollars. But the Civil War dealt river traffic a blow from which it could not recover. The temporary suspension of river commerce during hostilities had given the railroads an opportunity, of which they took full advantage, to gain rapidly in the Mississippi Valley the ascendancy they had won more slowly in the East. It may be said, therefore, that the Civil War ended the first period of waterway transportation.

In the second period, which lasted for about fifty years after the Civil War, waterway transportation seemed about to expire altogether as the railroads became the undisputed rulers of the transportation field. Yet even as the waterways were being pronounced moribund, developments were taking place that were to make their rejuvenation possible. For the great advances in science and engineering during this period made possible improvements in marine engines, the adaptation of the propeller to shallow-draft vessels, and the success of the early experiments in the substitution of towboats and barges for the packet. These developments prepared the way for the third, or modern, era of waterway transportation.

The impetus for this third period came from World War I. The heavy burden that fell upon the nation's transportation system between 1916 and 1918 led the federal government to give serious consideration to the possibility of increasing the system's capacity by more effective use of the waterways. As a result, in 1918 the U.S. Railroad Administration initiated barge operations on various waterways. Although this experiment proved costly, it did have the effect of renewing interest in the use of waterways and set in motion a train of events which subsequently proved that, in the new age of science and technology, inland

---

[7] Adapted from *Water Resource Activities in the United States: Future Needs for Navigation*, Committee Print No. 11. Select Committee on National Water Resources, U.S. Senate, 86 Cong. 2 sess. (1960), pp. 1–2.

water transportation possessed potentialities undreamed of in the golden age of the packet. It became clear that with modern marine engines and the great barges which they made possible, modernization of the waterways would lead to development of a low-cost addition to the nation's transportation system. Beginning with the report of the Inland Waterways Commission in 1908, the federal government undertook a systematic improvement of the waterways so that modern equipment could be used efficiently. Since that time many miles of waterway have been deepened to 9 feet or more. The third period of water transportation is, therefore, generally considered to extend from the early 1920's to the present.

Structure of the Industry[8]

Approximately seventeen hundred companies are engaged in commercial operations on the inland waters of the United States: 113 companies are certificated by the Interstate Commerce Commission (ICC) to provide service as regular route common carriers; 32 companies hold ICC permits to provide services under contracts with shippers; 1,150 companies engage in the transportation of commodities that are exempt from regulation under provisions of the Interstate Commerce Act; and about 400 companies engage in private transportation of their own commodities.

These companies operate approximately 14,000 dry cargo barges and scows with a total cargo capacity in excess of 14,000,000 tons; 2,600 tank barges with a total cargo capacity of approximately 5,150,000 tons; and 3,800 towboats and tugs with a total aggregate power in excess of 2,700,000 horsepower. About 80,000 persons are employed aboard the inland fleet. An equal number are employed in shore-based work directly connected with inland fleet operations—office personnel, terminal operators, service personnel, and shipbuilding and ship repair personnel.

Continuing technological advances will expand the opportunities for exploitation of the advantages of water transport. Some of the future improvements probably will come through co-ordination of services by the various modes and through some form of development of containerization adapted to combined service of inland water carriers with rail and highway operations.

Towboats

Towboats and tugboats form an indispensable team in U.S. transportation. Built to the most precise design specifications, and equipped with modern navigational instruments and safety devices, there is little to relate modern towboats to their ancestors—the stern- and side-wheeler steamboats that once sailed the rivers in great numbers.

A wide variety of towboats ply the nation's waterways today, ranging from vessels with single propellers to vessels with four propellers, each driven by an individual diesel engine. Sizes range from approximately 36 feet long, 12 feet wide, 6 feet draft, with engines of about 100 horsepower; to 170 feet long, 58 feet wide, a draft of 10 feet 3 inches, and four screws driven by engines that develop up to 9,000 horsepower. Towboats of 6,000 horsepower and up are capable of pushing barges carrying as much as 40,000 to 50,000 tons of cargo. In comparison, a modern diesel freight locomotive of 6,000 horsepower can efficiently handle a train of 120 cars loaded to an average of 50 tons per car, a total of approximately 6,000 tons.

The towboat is used for push-towing operations on most of the inland systems where the water routes are protected by surrounding land masses and where the waters are either

[8] The data on pp. 10–16, with the exception of the tables, are taken from *Big Load Afloat: U.S. Inland Water Transportation Resources* (Washington, D.C.: The American Waterways Operators, Inc., 1965).

TABLE 1. Vessels Transporting Freight on Inland Waterways of the United States,[a] December 31, 1964

Types of vessels	Mississippi River System	Atlantic, Gulf, & Pacific coasts waterways	Total
Self-propelled			
Towboats and tugs			
Number of vessels	1,864	2,001	3,865
Total horsepower	1,579,869	1,212,494	2,792,363
Not self-propelled			
Dry cargo barges and scows			
Number of vessels	10,590	3,849	14,439
Cargo capacity (net tons)	10,832,509	3,236,714	14,069,223
Tank barges			
Number of vessels	2,045	601	2,646
Cargo capacity (net tons)	4,180,257	977,064	5,157,321
Total not self-propelled			
Number of vessels	12,635	4,450	17,085
Cargo capacity (net tons)	15,012,766	4,213,778	19,226,544

[a] Exclusive of the Great Lakes.
Source: 1966 Inland Water-Borne Commerce Statistics (Washington: The American Waterways Operators, Inc., April 1968). Data originally from the U.S. Army, Corps of Engineers.

relatively calm in their natural state (as on the Lower Mississippi and the Missouri rivers), or where a system of locks and dams creates relative calmness. For push-towing, the barges are tied rigidly together by steel cables or ropes to form a single unit, and this unit is then lashed solidly against the boat's towing knees. The relatively flat-bottomed towboat with massive power in its propellers also has a set of multiple rudders which afford maximum control for the forward, backing, and flanking movements that are required to navigate the restricted channels of rivers and canals.

Character of the waterway, condition of the waterway, lockage conditions, size of tow, and horsepower of the towing vessel in relation to the size of tow all influence origin-to-destination running times of towboats. The following are typical transit times for the average tow over typical inland routes under ideal conditions:

- Pittsburgh–New Orleans, 1,852 miles: upstream 14 days and 2 hours, downstream 8 days and 18 hours;
- Kansas City–New Orleans, 1,434 miles: upstream 11 days and 22 hours, downstream 6 days;
- Minneapolis–New Orleans, 1,731 miles: upstream 13 days and 12 hours, downstream 7 days and 22 hours;
- Chicago–New Orleans, 1,418 miles: upstream 11 days and 8 hours, downstream 6 days and 7 hours;
- Pittsburgh–Houston (via New Orleans), 2,257 miles: upstream 16 days and 21 hours, downstream 10 days and 20 hours.

Power and Control

Almost all the towboats and tugboats in the United States are now powered by diesel engines. The steam-powered vessel is a relic of the past and only a very few are left in service. Once started, the shift from steam to diesel propulsion took place very rapidly. The diesel towboat is responsible for bringing the barge and towing vessel industry to its place of national prominence on the transportation scene in the last twenty-five years.

When the propeller replaced the stern wheel or side wheel on river and canal boats, an immediate problem was posed by the limitation imposed on the size of the propeller by the small space between the bottom of the boat and the bottom of the channel. This problem has been met in two ways: by development of more efficient propellers and by development and perfection in towboats of the tunnel stern. The tunnel stern is a design feature in which part of the propeller is actually above the level of the water surface in a spoon-shaped recess in the bottom of the hull that is filled with water by vacuum action when the propeller is turning. Another device that improves propeller efficiency in river towboats is the Kort nozzle, a funnel-shaped structure built around the propeller to concentrate the flow of water to the propeller. Under certain favorable operating conditions the Kort nozzle is reported to add as much as 25 per cent thrust to the propeller. A recent innovation, the bow steering unit, greatly increases maneuverability and, as a result, average speed, by applying sideward thrust at the bow of large tows. This innovation may be more important than the introduction of the Kort nozzle.

One of the most important gains in efficiency came from the development, principally during World War II, of dependable reversing-reduction gears capable of transmitting high horsepower. Before these gears were available, the conflict between high engine efficiency at high rpm and high propeller efficiency at low rpm usually resulted in a sacrifice of efficiency at both ends. Good reversing-reduction gears permit engine operation at the most efficient rpm.

The modern pilot has available both short- and long-range radiophones. The long-range radio is kept tuned in on the intership channel and consequently long before two approaching boats are within sight of each other or close enough to exchange passing signals by whistle, the two pilots have made radio contact, discussed the navigating conditions involved in their passing, and have agreed how they will pass. Fog, rain, and snow no longer mean a complete cessation of activities. Radar presents the pilot with a constant map of the river showing his position with relation to the shape of the river and any object in it.

Barges

The inland waterways industry has demonstrated that virtually any commodity can be shipped by water through development of a variety of types and sizes of barges for the efficient handling of products, ranging from coal in open hopper barges to chemicals in "thermos bottle" barges, and from dredged rock in dump scows to railroad cars on car-floats. Barging is the only practical mode for long distance moving of outside machinery, tanks, kilns, and some of the space vehicles.

Twenty-five years ago most river barges were designed as single individual units, with a rake, or slope, on each end. For navigating singly, this form is still most efficient. However, model testing showed that the assembly of multiple units of this form in a single tow resulted in great loss of efficiency by the cumulative drag of many water-breaking rakes in the middle of the tow.

Some barges are now designed to be assembled into integrated tows that will have an underwater shape and water resistance that is nearly the equivalent to that of a single vessel. Such an integrated assembly has a lead barge with an easy rake at the bow, to minimize the resistance of the water, and a square stern. The trailing barge has a short rake on the stern, and a square bow. Between the lead barge and the trailing barge, double square-ended barges are inserted, thus eliminating any underwater surface break. An additional benefit from this kind of integrated tow is the increase in capacity caused by the added buoyancy of the square ends of the barges.

The integrated high-speed tow is generally efficient for the carriage of a large volume of a single commodity over a long distance on a continuing basis. Identical draft of all barges comprising the tow is vital to the efficiency of the operation. The transportation of petroleum and petroleum products, chemicals, and other liquids for which tows can continuously operate as a unit has been most successful.

Make-up of the tow is sometimes handled by a special crew using a small workboat. In many of the major ports this work—as well as the break-up of tows and distribution of barges to loading and unloading docks—is handled by harbor fleet operators. Thus line-haul towboats can arrive in port, cut loose from their tow of barges, take on another tow which has already been made up, and be on their way again in an hour or two.

The Channels

The United States has 25,380 miles of navigable inland channels, exclusive of the Great Lakes, of which 15,348 miles have a depth of 9 feet or more. Except for the 522-mile New York State Barge Canal, all of these waterways are federal projects.

With the exception of the Upper Mississippi Waterway, the Missouri River, and the New York State Barge Canal, all of which are closed by ice from December through March, the inland channels are open to navigation throughout the year. At times, ice forms on the Illinois Waterway, the Mississippi above St. Louis, and on the Ohio River, but navigation is seldom impeded for any length of time.

With two notable exceptions, the channels are slack water routes which have been improved for navigation by the construction of systems of locks and dams. The Mississippi

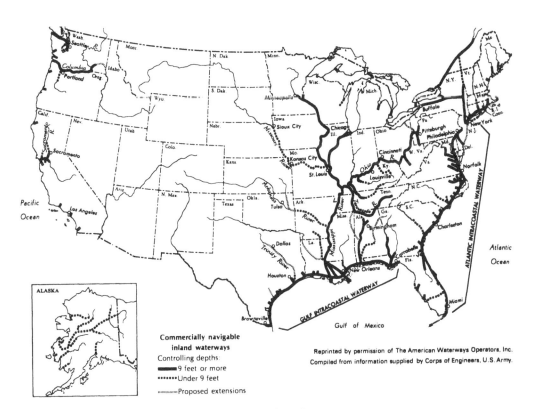

FIGURE 2. Map of inland waterways.

is open river for 1,000 miles south of St. Louis. The Missouri is open river. Yet the two present a striking difference. The Mississippi is a wide, deep, commodious river. The Missouri has a restricted 7-foot depth. Both the Atlantic and Gulf Intracoastal Waterways are largely open channels, although both have some locks and both have reaches that are exposed to tidal currents and winds.

Much of the current waterway construction involves work to widen or deepen channels or to modernize channels by the construction of higher dams and larger locks. The redevelopment of the Ohio River, for instance, is under way. The river was canalized in 1929 by a system of fifty-two locks and dams. The new canalization project started in 1955 will replace the present lock and dam system with nineteen new high-lift locks with 110- by 1,200-foot lock chambers. The present chambers are 110 feet by 600 feet.

Recanalization of the Warrior-Tombigbee river system in Alabama is under way. The Corps of Engineers has completed work on a study to build a second system of locks on the Illinois Waterway. The widening and deepening of the Gulf Intracoastal Waterway has been authorized and is under way. Water resource development interests along the Upper Mississippi River are urging studies of the possibility of recanalization with higher and fewer dams and bigger lock chambers. These projects will not add any channel mileage.

Some water resource development interests are seeking additional navigable channels. The Tennessee-Tombigbee Waterway has been authorized for improvement by the Congress but no funds have been made available. This project would connect the Tom-

TABLE 2. Commercially Navigable Inland Waterways of the United States by Lengths and Depths[a]

(miles of waterways)

Group	Under 6 ft	6 ft to 9 ft	9 ft to 12 ft	12 ft to 14 ft	14 ft and over	Total
Atlantic Coast Waterways (exclusive of Atlantic Intracoastal Waterway from Norfolk, Va. to Key West, Fla.)	1,502	1,271	593	975	1,490	5,831
Atlantic Intracoastal Waterway from Norfolk, Va. to Key West, Fla.	–	211	65	954	–	1,230
Gulf Coast Waterways (exclusive of the Gulf Intracoastal Waterway)	2,048	718	1,239	216	444	4,665
Gulf Intracoastal Waterway from St. Marks, Fla. to the Mexican border, including Port Allen–Morgan City alternate route	–	–	–	1,177	–	1,177
Mississippi River System	2,400	684	4,449	732	273	8,538
Pacific Coast Waterways	725	370	239	26	2,182	3,542
All other waterways	45	58	–	8	286	397
Total	6,720	3,312	6,585	4,088	4,675	25,380

[a] The mileages in this table represent the lengths of all navigable inland channels of the United States (exclusive of the Great Lakes), including those improved by the federal government, other agencies, and those which have not been improved but are usable for commercial navigation.

Source: 1966 Inland Water-Borne Commerce Statistics, p. 1.

bigbee River from its headwaters in Mississippi with the Tennessee River at Pickwick Dam, thus providing a route from the Tennessee River, through Mississippi and Alabama, to connect with the Gulf Intracoastal Waterway at Mobile. A proposal has been made to extend the Gulf Intracoastal Waterway from its eastern terminus at Carrabelle, Florida, along the shoreline of Florida to Tampa. Construction has begun on the Cross-Florida Barge Canal, a 185 mile, 12-foot deep waterway with five single-lift locks, 84 feet by 600 feet. It will provide a link from the Atlantic Intracoastal Waterway at Jacksonville, Florida, to the Gulf of Mexico through the Withlacoochee River valley 95 miles north of Tampa on the west coast of Florida.

Construction of nineteen locks and dams on the Arkansas-Verdigris river system is now under way and will provide an additional 450 miles of navigable channel from the Mississippi River to Catoosa, near Tulsa, Oklahoma, when completed in 1970.

By 1975, a 9- by 300-foot channel should be completed on the Missouri River from its confluence with the Mississippi just north of St. Louis, Missouri, to Sioux City, Iowa. The Missouri from its mouth to Kansas City now has a controlling dimension of 7½ feet by 250 feet; from Kansas City to Omaha, of 8 feet by 220 feet; and from Omaha to Sioux City, of 8½ feet by 250 feet.[9]

Construction work is under way to canalize the Alabama River from its confluence with the Mobile River just north of Mobile, Alabama, to Montgomery, with prospects of extending the navigable system to Gadsden, Alabama.

The Chattahoochee River has been canalized and is open to navigation as far north as Columbus, Georgia, and studies are under way on extending this navigation system to Atlanta.

Completion of the initial construction phase of a waterway does not end the need for dredging. Normal silting builds up shoals and bars. Vessel movements themselves sometimes shift channel bottoms and contour lines. High water flows also increase silting and cause shoals and bars to build up. To maintain the navigation channels, supervised dredging is necessary on most waterways. It is done under direction of the Corps of Engineers, primarily by private contractors.

Locks, Dams, and Terminals

The sizes of the lock chambers which pass vessels from one level of water to the other in canalized streams have tended to dictate standardization of the dimensions of vessels using the inland channels.

A common tow configuration for passage through the large 110- by 1,200-foot lock chambers would be four barges wide and six barges long (using 175- by 26-foot barges). For passage through the small 110- by 600-foot lock chambers, a common configuration is four barges wide by three long. Jumbo barges (195 feet by 35 feet) are frequently made up in other configurations compatible with lock chamber size.

Lock chambers of adequate size to accommodate the type of tows that operate on a waterway are important to the economics of barge transportation. The average lock is designed to accommodate the passage of vessels in a 20- to 30-minute operation. Single tows which are too large to pass through a lock in a single operation require double lockage. Break-up and reassembly of the tow plus the two lockage operations takes about an hour and a half. Since operating costs of a towboat range from $50 to $100 per hour, double lockages impose a cost penalty to operators and added costs to shippers.

The major portion of terminaling work to service inland water carrier operations is carried out through facilities provided by shippers and receivers. Modern terminals are highly mechanized installations designed to permit fast loading and delivery, thereby

[9] This paragraph was amended in keeping with changes suggested by the Missouri River Division, U.S. Army Corps of Engineers, to update progress on the river.

Corps of Engineers Photo

FIGURE 3.   Markland Locks and Dam, U.S. Army Engineer District, Louisville, Kentucky.

eliminating unnecessary delays and speeding up vessel turn-around time. Mechanization is, of course, readily applicable to the handling of bulk-loading cargoes at low cost. Terminaling costs on non-bulk freight are necessarily higher because of the labor involved.

The U.S. Army Corps of Engineers

For almost a century and a half, the Corps of Engineers has been responsible, under congressional authorization, for planning, constructing, maintaining, improving, and operating inland waterways, including harbors, for commercial navigation in the United States. This responsibility includes engineering feasibility studies, cost studies, economic analysis, and development of overall justification data as a basis for congressional action to authorize and finance river and harbor improvements.

In addition, the Corps has full engineering and construction management responsibility for the following work:

- Providing and maintaining channels at their authorized depth and width;
- Improving and maintaining harbors, including provision of protective works such as jetties and breakwaters;

- Providing means other than lighting and marking of channels for facilitating navigation;
- Canalization where locks and dams are required;
- Removal of obstructions to maintain the navigability of the waterways.

The Corps' Board of Engineers for Rivers and Harbors was established by Section 3 of the River and Harbor Act of June 13, 1902. The Board consists of a senior member, generally the Deputy Chief of Engineers; five officers who are usually Division Engineers assigned to Board membership in addition to their other duties; and a resident member permanently stationed for his tour of duty with the Board in Washington, D.C. It reviews all reports on authorized preliminary examinations and surveys of river and harbor projects and of flood control projects and reports its conclusions and recommendations to the Chief of Engineers, who forwards the report with his own conclusions and recommendations to the Secretary of the Army for transmittal to Congress. The Board is also authorized, when requested by the appropriate committees of Congress, to review and report through the Chief of Engineers upon any authorized project or desired improvement with a view to recommending the initiation of a project or modification of an existing project. The Chief of Engineers may refer to the Board for consideration and recommendation all special reports ordered by Congress and may prescribe such other duties as he desires.

The Comparative Position of Inland Waterway Transportation and Its Commodity Composition

Tables 3 and 4 give a brief picture of the relative size of the inland waterway transportation industry and of the commodity mix of the industry.

Public Investment in Inland Waterways

The Corps of Engineers received its first appropriation for river improvement in 1824. Appropriations from then until 1959 for the construction, operation, and maintenance of navigation works amounted to approximately 5.5 billion dollars.[10] Since that time, authorizations have amounted to nearly that amount again, including approximately one billion dollars each for the Ohio, Arkansas, and Trinity rivers.

Federal investment in the various parts of the inland waterway system varies greatly, depending on the physical characteristics of the waterway and the period of the investment. Figure 4, though substantially out of date, shows the relative levels of investment for the provision of standard channels for navigation. The widely varying costs of channel improvements are clearly exhibited. The investment profile indicates that the level of investment becomes heavier as the waterways are developed away from the main alluvial rivers. Traffic density, with some exceptions caused by high levels of local traffic, tends to taper off inversely with the level of investment. In this way, points are reached on all waterways beyond which navigation improvements can not be economically justified.

[10] *Water Resources Activities in the United States: Future Needs for Navigation*, Committee Print No. 11, Select Committee on National Water Resources, U.S. Senate, 86 Cong. 2 sess. (1960).

TABLE 3. Freight Traffic in the United States, 1960–1965

Amount and mode	1960	1961	1962	1963	1964	1965
Net tons (millions)						
Railways[a]	1,241 (45.0%)	1,194 (43.9%)	1,234 (43.2%)	1,285 (43.4%)	1,356 (43.3%)	1,388
Motor trucks	276 (10.0)	289 (10.6)	321 (11.2)	338 (11.4)	366 (11.7)	(b)
Great Lakes	155 (5.6)	137 (5.0)	136 (4.8)	142 (4.8)	151 (4.8)	154
Inland waterways	395 (14.3)	388 (14.3)	418 (14.6)	431 (14.5)	457 (14.6)	472
Pipelines	692 (25.1)	712 (26.2)	749 (26.2)	766 (25.9)	801 (25.6)	839
Total	2,759	2,720	2,858	2,962	3,131	–
Ton-miles (billions)						
Railways[a]	575 (43.4)	570 (43.2)	600 (43.1)	629 (43.1)	666 (43.4)	709 (42.6)
Motor trucks	299 (22.5)	305 (23.2)	331 (23.8)	348 (23.8)	350 (22.8)	388 (23.3)
Great Lakes	99 (7.5)	87 (6.6)	90 (6.4)	95 (6.5)	106 (6.9)	110 (6.6)
Inland waterways	124 (9.4)	123 (9.3)	133 (9.6)	139 (9.5)	144 (9.4)	153 (9.2)
Pipelines	229 (17.2)	233 (17.7)	238 (17.1)	250 (17.1)	269 (17.5)	306 (18.3)
Total	1,326	1,318	1,392	1,461	1,535	1,666

Note: Percentages of total shown in parentheses.
[a] Mail and express traffic has been eliminated from railroad totals.
[b] Tonnage figures for motor trucks not available for 1965.
Source: 1966 Inland Water-Borne Commerce Statistics, p. 6. Original data from annual reports of the Interstate Commerce Commission.

TABLE 4.   Principal Commodities Transported on the Inland Waterways of the United States (exclusive of the Great Lakes), 1966

Commodity	Net tons (2,000 lb)
Grain and grain products	16,773,062
Soybeans	3,672,140
Fresh fish and shellfish	1,738,162
Marine shells, unmanufactured	24,421,773
Iron ore and concentrates	2,085,326
Bauxite and other aluminum ores and concentrates	503,016
Bituminous coal and lignite	110,149,626
Crude petroleum	49,306,628
Limestone flux and calcareous stone	4,450,131
Crushed and broken stone	8,306,756
Sand and gravel	52,619,049
Clay, ceramic, and refractory materials	2,205,668
Sulphur, dry and liquid	5,544,812
Nonmetallic minerals, excluding fuels, n.e.c.	2,877,808
Sugar	868,817
Molasses, inedible	770,621
Rafted logs	22,638,240
Pulpwood	1,722,808
Lumber and lumber products	1,802,886
Paper and paper products	1,584,893
Sodium hydroxide	2,676,756
Crude products from coal tar, petroleum, and natural gas	2,060,782
Alcohols	1,552,663
Sulphuric acid	3,450,668
Benzene	1,395,192
Basic chemicals and products, n.e.c.	6,261,276
Fertilizer and fertilizer materials	1,737,782
Miscellaneous chemical products	1,778,857
Gasoline	38,796,539
Jet fuel	6,457,172
Kerosene	4,006,623
Other petroleum and coal products	67,244,714
Building cement	3,729,421
Iron and steel products	5,862,691
Iron and steel scrap	1,059,796
Total—principal commodities	462,113,154
All other commodities	26,953,056
Grand total	489,066,210

Source: 1966 Inland Water-Borne Commerce Statistics, p. 5. Original data from U.S. Army Corps of Engineers.

19

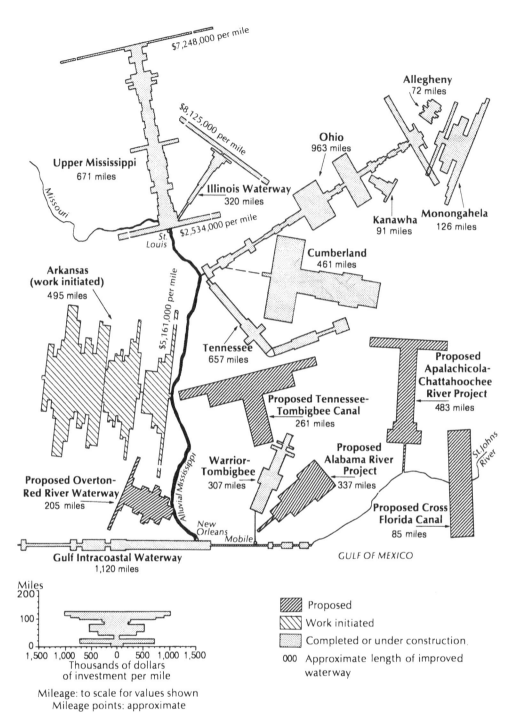

$7,248,000 per mile

Allegheny
72 miles

$8,125,000 per mile

Ohio
963 miles

Upper Mississippi
671 miles

Missouri

Illinois Waterway
320 miles

Kanawha
91 miles

Monongahela
126 miles

$2,534,000 per mile

St.
Louis

Cumberland
461 miles

Arkansas
(work initiated)
495 miles

$5,161,000 per mile

Tennessee
657 miles

Proposed
Apalachicola-
Chattahoochee
River Project
483 miles

Proposed Tennessee-
Tombigbee Canal
261 miles

St. Johns
River

Proposed
Alabama River
Project
337 miles

Warrior-
Tombigbee
307 miles

Proposed Overton-
Red River Waterway
205 miles

Alluvial Mississippi

New
Orleans
Mobile

Proposed Cross
Florida Canal
85 miles

Gulf Intracoastal Waterway
1,120 miles

GULF OF MEXICO

Miles
200
100
0
1,500 1,000 500   0   500 1,000 1,500
Thousands of dollars
of investment per mile

Proposed

Work initiated

Completed or under construction.

000   Approximate length of improved
waterway

Mileage: to scale for values shown
Mileage points: approximate

Note: Presentation of investment data in this chart is illustrated above (left). Mileages are scaled along the schematic course of the waterway. Average investment per mile for waterway sections is scaled at right angles to the course of the waterway. For example, from the mouth of the Monongahela to mile 11.2, the investment per mile is $243,000; from mile 11.2 to mile 23.8, the investment per mile is $1,418,000.

FIGURE 4.   Distribution of original navigation investment expenditures on the Mississippi River System and adjacent waterways. *Source:* U.S. Department of Commerce, *User Charges on Inland Waterways* (January 1959). Originally prepared in the Office of the Under Secretary of Commerce for Transportation, Department of Commerce, from data contained in the U.S. Army Corps of Engineers, *Annual Report for the Fiscal Year Ended June 30, 1957*, Vol. 2, and the *Annual Report of the Tennessee Valley Authority for the Fiscal Year Ended June 30, 1957.*

B. NAVIGATION TRADE-OFFS ON THE MISSOURI RIVER

The following paragraphs have been taken from a very informative paper by T. Waara concerning integrated operation of Missouri River reservoirs.[11] They are included here in order to illustrate clearly the types of trade-offs which exist among the outputs of a multiple purpose river system. In particular, present interest centers on trade-offs between navigation, electrical power generation, irrigation, flood control, and recreation. The opportunity costs of water released for the maintenance of navigation represent a consideration which has been omitted from the studies of this volume. While these trade-offs are perhaps most clearly seen on the Missouri, they are present (although quantitatively less important) on the Columbia, Upper Mississippi, and Ohio rivers.

The watershed of the Missouri River, constituting more than one-seventh of the land mass of the 48 contiguous United States, offered for many decades a challenging example of wasteful and often ravaging neglect of water resources. These resources are now being regulated and put to constructive use by six Missouri River dams with a gross storage in excess of 94,000,000,000 cubic meters. Compared to a mean flow past the lowermost dam of almost 30,000,000,000 cubic meters per year, this high ratio of storage to flow has provided much operational flexibility in regulating the system.

The fact that choices were possible led to sharp differences of opinion during early days of system operation concerning appropriate objectives. These differences, which were accentuated by a severe drouth during the initial-fill period, centered around whether navigation or power generation should receive preferential treatment. . . .

The marketing area for Missouri Basin power experiences its peak load during the winter months, usually in December. Since ability to fulfil contracts for firm power rests upon ability to meet peak loads, it was desirable from the standpoint of power production to operate the main stem projects in such a way as to hold winter generation at a relatively high level in reference to summer loads. Any sizeable summer excess had to be marketed as dump power at a considerably lower price. And since the output of a hydro plant increases with increasing head, power interests placed a premium upon early and rapid storage gain.

Navigation, on the other hand, required the maintenance of adequate channel depths during the open water seasons; cutting winter flows to minimum sanitation levels was a means its proponents suggested to conserve the water supply during those lean years. Here assuredly was an urgent and knotty requirement to provide the unity of purpose needed in administering the young system. . . .

This flexibility [provided by the adequate storage reserves now on hand] has provided a solution to the early conflict between the requirements of service to power and service to navigation. It will be recalled that navigation needs summer flows of two or three times the magnitude necessary for minimum water supply and stream sanitation levels in the winter. The market for firm power, on the contrary, reaches a peak in the winter months and any summer generation greatly in excess of this level must be sold at reduced dump power rates. Resolution of this seeming incompatibility was accomplished by operating the system as two separate and complementary parts. The lower two reservoirs, Fort Randall and Gavins Point, follow a pattern of releases designed to serve navigation. . . .

Generation of electrical energy at Fort Randall and Gavins Point follows, perforce, the release pattern. For somewhat over half the navigation season Oahe, the next upstream storage project, also follows this pattern of high releases, backing up the outflow from the downstream projects which contain relatively little conservation storage when compared to the nearly 48×10^9 m³ (39 million acre-feet) shared by the three upstream reservoirs. (Big Bend which lies between Oahe and Fort Randall is essentially a run-of-river plant; its releases parallel those from Oahe except for some daily and weekly ponding.)

After Labor Day, however, Oahe releases are cut back and the drawdown of Fort Randall begins. By this means nearly 2,466 million m³ (two million acre-feet) of conservation storage is

[11] T. Waara, "Integrated Operation of Missouri River Reservoirs for Multiple-Purpose Use," *Proceedings of the Sixth Congress of the International Commission on Irrigation and Drainage* (New Delhi, India, 1966), pp. 22.25–22.44.

evacuated from Fort Randall prior to the close of navigation. After that date, when Fort Randall releases fall to the vicinity of 227 m³/sec. (8,000 cfs), Oahe and Big Bend can operate throughout the winter with releases in the 425 m³/sec. (15,000 cfs) range, refilling the Fort Randall pool and generating additional winter power.

While this pattern of high summer releases and low winter releases still prevails in the lower part of the system, an opposite role is assigned to the upper portion. During the planning phase, channel capacity of the Missouri River under ice cover was set at 283 m³/sec. (10,000 cfs) below Fort Peck and 425 m³/sec. (15,000 cfs) below Garrison, for want of actual experience at higher flows.

Several years of repeated testing have now shown that the Missouri River channel below Fort Peck can safely carry 354 m³/sec. (12,500 cubic feet per second) under ice cover. At Garrison 779 m³/sec. (27,500 cfs) have been released without difficulty once the channel is solidly frozen in. These high levels of discharge and the associated high generation rate are utilized to offset the drop in energy which winter brings to the lower system. When navigation starts and the downstream projects begin their season of heavy generation, Fort Peck and Garrison are cut back to a level which avoids the generation of large amounts of very cheap dump power which once seemed a threat to the economic structure of the projects. . . . Even the load requirements themselves are changing in a manner which eases accomplishment of balanced operation. Although the peak load for the marketing area as a whole still falls in December, the growth of irrigation and airconditioning loads is being reflected in a steady increase in the July and August demand.

This is not to suggest that dump power will become a thing of the past. An unbalanced distribution of runoff into the system, such as was experienced just in 1964, places heavy pressures quite apart from the seasonal release pattern, to adjust storage at the possible price of some generation of dump power. The long-range prospect, however, with construction of the system near completion and firm power commitments approaching full development, is for a concurrence and not a conflict of interests between power and navigation.

The resolution of an important conflict between two primary functions of the system has not brought to an end all objection to the shape of things as they are, accompanied by pointed suggestions on how operations could be improved. Irrigators pumping directly from the channel are quick to react when a reduction in release cuts off their inlets behind newly forming sandbars, even though total flow in the river is several thousand times greater than the need. Increased discharges which might inundate pump motors are equally unwelcome. Ferry operators call for more releases when their boats go aground or approach to access sites becomes difficult. Shippers have requested special consideration of an extension of navigation into the period of usual winter freeze-up, and recreational interests call for both constant pool levels and constant discharges, while contractors building bridges and industrial plants submit detailed requests for control of stage and flow. These are earnest and well-meaning people who recognize a condition less than ideal in their locality or field of interest and also recognize how it might be corrected. Not so apparent to them is the consequence these corrective measures might have upon the intricate interrelationships of a system wherein power revenues alone average $60,000 a day and the value of stored water has been estimated at $1.60 per 1,000 cubic meters (two dollars an acre-foot). When one of these special requests can be accommodated, however, without unduly burdening the primary functions, its scheduling and coordination are carried out by the Reservoir Control Center.

From the above, it is seen that the provision of vast amounts of storage and the optimizing of operating procedures for the system have served to reduce the sharpness of the trade-offs between navigation and other outputs on the Missouri system; i.e., the operating costs of navigation in terms of the other outputs have been reduced through investment in reservoir storage. The extent to which storage should be developed is part of the problem of optimum river system design. Opportunity costs remain, whatever the degree of regulation of the river, and should be taken into account in deciding upon optimum operating procedures for the river system.

2

TECHNOLOGY AND PRODUCTION FUNCTIONS
FOR THE TOW

The bargeline industry is regulated with respect both to the routes which may be served by the various common-carrier firms and to tariffs charged on some of the common-carrier operations. Companies which control trucking firms or railroads are not permitted to own bargelines. These regulations (and presumably the exemptions from them) should be based on considerations of efficiency in operations and of equity among carriers and their customers; the latter directed mostly toward the avoidance of uncontrolled monopoly situations. Whether or not such constraints are in the public interest depends upon the technology of inland waterway transport (both the public and private parts of the system) and the extent to which efficiency is related to firm size. Ultimately, the degree of potential competition within the bargeline industry and between barge and rail depends upon these technological factors in the two industries. In order to evaluate the current regulatory structure as well as forecast future modal splits of freight traffic, it is necessary to know the extent to which the efficiency of operations increases with size—the size of tow, the size of the firms in the bargeline industry, and the size (length and cross-sectional characteristics) of the inland waterway system.

If larger tows are more efficient, firms which command a larger share of the market will have cost advantages over smaller firms because they are able to schedule larger tows. If larger firms experience increasing efficiency with size, there will be a tendency toward the concentration of business in the hands of a few large firms—a condition possibly requiring regulation. Whether or not such regulation is needed depends on the degree of competition with other modes of transport which, in turn, in the long run depends on what happens to efficiency and costs as the entire bargeline industry expands.

This chapter and Chapter 3 present the technology of the industry at two levels: (1) the individual transport unit—the tow;[1] and (2) the firm. The technology of the tow is considered here in order to determine the extent to which tows are subject to economies of scale and how they are affected by the characteristics of the waterways in which they travel. Such information is useful to bargeline firms in the selection of optimum equipment for particular types of haul and to the analyst who seeks to explain why firms exhibit economies or diseconomies of scale.

---

[1] A *tow* consists of a *flotilla* of barges rigidly lashed together and pushed by a *towboat*. Equipment types vary greatly. The data used in this study refer to the equipment which is considered standard for the industry and is found most frequently in service; namely, diesel-powered towboats with Kort nozzles (funnel-shaped shields around the screw which produce a better flow of water to the screw and reduce cavitation, thereby increasing efficiency), and 195- by 35-foot barges.

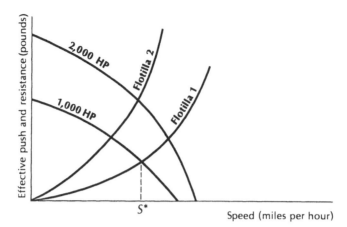

FIGURE 5.   Determination of equilibrium speed.

The major aspects of towing technology have been presented elsewhere [Howe, Jan. 1964, July 1964, Nov. 1967; Horton, 1958; Kelso, 1960]. The following factors must be taken into consideration in deriving empirical functional representations of tow performance:

- The resistance of the barge flotilla as a function of speed, and characteristics such as length, breadth, and draft, etc.;
- The push generated by the towboat as a function of the size of boat (rated horsepower) and its speed;
- The effects of the environment on both of the above factors; in particular, the effects of depth, width, and current of the waterway in which the tow operates.

The basic technological constraint which must hold for any tow proceeding at constant speed is:[2]

$$EP = R. \tag{1}$$

If EP can be related to towboat characteristics and speed and R to flotilla characteristics and speed, then equation (1) will permit determination of the equilibrium speed for any boat-flotilla combination as indicated by Figure 5. Since it is known that depth and width of waterway will affect both flotilla resistance and towboat effective push, if depth and width can be directly incorpo-

---

[2] The following notation is used:
A = total deck area of the barge flotilla, in square feet;
B = overall breadth of barge flotilla, in feet;
D = depth of waterway, in feet;
EP = effective push of towboat, in pounds force, as a function of its rated brake horsepower (HP);
H = draft of barge flotilla (assumed uniform), in feet;
L = overall length of barge flotilla, in feet;
R = resistance of barge flotilla, in pounds force;
R_d = slope-drag force, in pounds force;
S = speed, in mph (still water);
W = width of waterway, in feet.

rated in the R and EP functions, the effects (on overall tow performance) of waterway improvements that involve increased depth and/or width can be evaluated directly.

Extensive investigation [Howe, Jan. 1964, Nov. 1967] has shown that the resistance of the barge flotilla can be represented by[3]

$$R = 0.07289 \; e^{\left(\frac{1.46}{D-H}\right)} \; S^{2.0} \; H^{0.6+\left(\frac{50}{W-B}\right)} \; L^{0.38} \; B^{1.19} \tag{2}$$

and the effective push of the towboat by

$$EP = 31.82 \; HP - 0.0039HP^2 + 0.38 \; HP \; D - 172.05S^2 - 1.14S - HP . \tag{3}$$

When a tow is traveling up a flowing river, not only must it overcome the resistance to passage through the water but it must also carry its burden uphill. This additional force is known as the slope-drag force, R_d, and must be added to or subtracted from the resistance as given in equation (2), depending upon the direction of travel. R_d can be expressed as a function of stream current, channel depth, and tow displacement. Equilibrium tow speed relative to the water, S^*, can then be determined by solving the equation

$$EP \; (S^*, \; \ldots) = R \; (S^*, \; \ldots) + (-1)^{\delta+1} \; R_d ; \tag{4}$$

when $\delta = 1$ for upstream travel and $\delta = 0$ for downstream travel. Tow speed over the ground can then be determined as

$$\hat{S} = S^* + (-1)^\delta \; S_w ; \tag{5}$$

when S_w is stream current. Thus, equilibrium speed is determined as a function of characteristics of the tow (HP, L, B, H) and characteristics of the waterway (D, W, S_w), plus the direction of travel, δ. Fixing these variables determines not only the tow's speed but, for a particular type of barge, the net cargo tonnage of the tow, $T(L, B, H)$. Then, by definition, the *rate* of output of the tow in net cargo ton-miles per hour is given by:

$$TM \equiv \hat{S} \; (HP, L, B, H; D, W, S_w) \cdot T \; (L, B, H) . \tag{6}$$

This, then, is the production function for the tow, as conditioned by the characteristics of the channel in which it operates. The function is quite complicated and can be analyzed only through numerical evaluation.[4]

NUMERICAL EVALUATION OF THE PRODUCTION FUNCTION

As an example of the characteristics of this production function, the following tabular and graphical presentations are based on the assumptions of a waterway 200 feet wide by 12 feet deep, a flotilla draft of 8.5 feet, and a flotilla length-to-breadth ratio of 5.57 (the ratio for a single 195- by 35-foot barge). Keeping

---

[3] The formula assumes a uniform draft, H, for all barges in the flotilla. Extensive numerical testing of the derived speed function (5) has indicated that the use of average draft for flotillas of nonuniform draft produces better prediction than the use of maximum draft.

[4] For a more detailed discussion and empirical testing of the speed function, see Howe [Nov. 1967]. For more detailed characterization of the production function (6), see Howe [Jan. 1964].

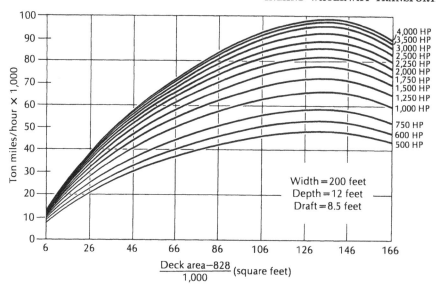

FIGURE 6.   Total product curves.

the length-to-breadth ratio constant, it is possible to measure the barge input in terms of total deck area of the barge flotilla, A. Figure 6 presents the total product curves for various towboat horsepowers as functions of the barge input.[5]

It is clear from Figure 6 that the production function for the tow exhibits normal attributes: (1) the total product curves for a fixed horsepower are strictly concave, reflecting the diminishing marginal productivity of the barge input; and (2) the marginal productivity of the horsepower input decreases. However, in the case of the towing production function, the marginal products of both inputs ultimately become negative. Further, beyond some point, the tow is subject to decreasing returns to scale *within a waterway of given depth and width.* This must always be true, for as flotilla width approaches channel width or as flotilla draft approaches channel depth, resistance increases without bound. Further, as the horsepower increases relative to the unobstructed channel cross-section,[6] effective push decreases because of restricted water flow to the screws, and resistance increases because of an extreme drawing of water from under the flotilla, causing the barges to "squat."

The marginal product schedules of the barge and horsepower inputs are illustrated for particular cases in the Tables 5 to 7. (See Chapter 2, footnote 2 for notations.)

Also of interest are the substitution possibilities which exist between the barge and horsepower inputs. There are, in fact, four major inputs for the individual tow: HP, A, labor, and fuel. For the individual tow, however, labor and fuel are directly related to the HP input and are not capable of independent

[5] One 195- by 35-foot barge has a deck area of 6,825 square feet. A four-barge flotilla has a deck area of 27,300 square feet; a nine-barge flotilla, 61,425 square feet; and a sixteen-barge flotilla, 109,200 square feet.

[6] Channel cross-section less the cross-sectional area of the flotilla.

TABLE 5. An Illustrative Marginal Productivity Schedule for the Barge Input[a]

Flotilla deck area (sq. feet)	Ton miles/hour	ΔTon miles/hour
6,825	13,030	–
16,825	27,149	14,119
26,825	38,653	11,504
36,825	48,579	9,926
46,825	57,356	8,777
56,825	65,201	7,845
66,825	72,223	7,022
76,825	78,499	6,276
86,825	84,011	5,512
96,825	88,807	4,796
106,825	92,832	4,025
116,825	95,911	3,079
126,825	97,989	2,078
136,825	98,817	828
146,825	98,126	−691
156,825	95,412	−2,714
166,825	90,153	−5,259

[a] $W = 200$; $D = 12.0$; $H = 8.5$; $HP = 4,000$; $L/B = 5.5714$.

TABLE 6. An Illustrative Marginal Productivity Schedule for the Towboat (Horsepower) Input[a]

Horsepower	Ton miles/hour	ΔTon miles/hour
500	46,683	–
1,000	63,024	16,341
1,500	73,756	10,732
2,000	81,299	7,543
2,500	86,594	5,295
3,000	90,118	3,524
3,500	92,144	2,026
4,000	92,832	688
4,500	92,276	−556
5,000	90,520	−1,756

[a] $W = 200$; $D = 12.0$; $H = 8.5$; $B = 138.5$; $A = 106,825$.

TABLE 7. An Illustrative Marginal Productivity Schedule for the Towboat Input with a Larger Flotilla[a]

Horsepower	Ton miles/hour	ΔTon miles/hour
500	43,517	–
1,000	59,169	15,652
1,500	69,659	10,490
2,000	77,209	7,550
2,500	82,682	5,473
3,000	86,515	3,833
3,500	88,957	2,442
4,000	90,153	1,196
4,500	90,186	33
5,000	89,092	−1,094

[a] $W = 200$; $D = 12.0$; $H = 8.5$; $B = 173.0$; $A = 166,825$.

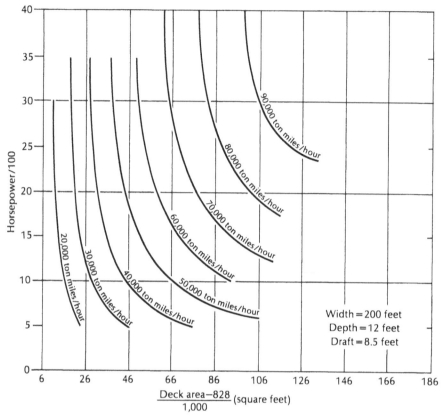

FIGURE 7.   Production isoquants for waterway transport.

variation. Thus, the substitution of HP for A really involves a substitution of HP, labor, and fuel for A. Figure 7 exhibits a family of isoquants for a fixed environment and draft. Obvious substitution possibilities exist but are extremely limited for low rates of output. The isoquants exhibit the usual convexity, and their spacing is indicative of "decreasing returns to scale" within a given operating environment. Actually, the isoquants do not have the usually assumed asymptotic properties since the marginal products of the inputs can become negative. The expansion paths associated with Figure 7 would be approximately linear, although the production function (6) itself does not appear to be homogeneous.

COST FUNCTIONS FOR THE TOW

If assumptions are made about boat and barge operating costs, the production function (6) may be used to trace out a family of average cost curves which define the envelope curve of average costs per ton-mile of output for the tow. For illustrative purposes, the following assumptions are made:

• Barge type is open hopper with a full cost per day (including depreciation) of $17.50;

- A tow will, on the average, spend 80 per cent of its time (19.2 hours/day in commission) running, 20 per cent in navigation and port delays;
- Stream current is 3 mph; channel width, 200 feet; and channel depth, 12 feet;
- Barges are loaded to a uniform 8.5-foot draft;
- The cost per day of operating a boat, as a function of boat size, is approximated by the following schedule:

Horsepower	Cost per day
500	$ 750
1,000	900
1,500	1,050
2,000	1,200
2,500	1,330
3,000	1,480
3,500	1,625
4,000	1,775
4,500	1,920

Of course, if conditions differ significantly from these assumptions, any other set of values could be substituted. These costs are intended to approximate full line-haul costs; i.e., all towboat costs, including depreciation and interest, labor, fuel, and miscellaneous, and all barge costs. They include no distribution of administrative overhead. Figure 8 illustrates the resultant cost curves. It

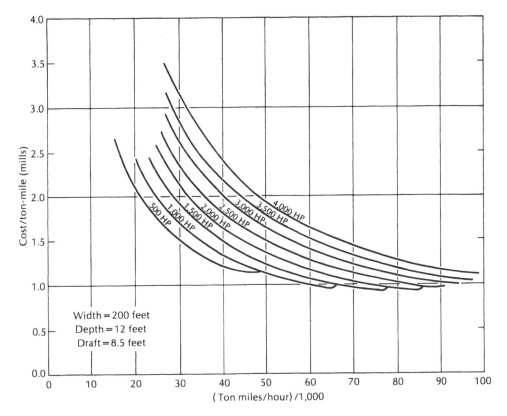

FIGURE 8.   Family of cost functions for waterway transport.

TABLE 8.   Joint Effects of Channel Width and Depth on Line-Haul Costs per Ton-Mile[a]

(mills)

Depths (feet)	Widths (feet)								
	150	200	250	300	350	400	450	500	550
12	23.71	1.21	1.07	1.03	1.00	0.99	0.98	0.98	0.97
18	19.64	1.02	0.90	0.87	0.85	0.84	0.83	0.83	0.82
24	18.11	0.94	0.83	0.80	0.78	0.77	0.77	0.76	0.76
30	17.05	0.88	0.78	0.75	0.73	0.72	0.72	0.71	0.71
36	16.22	0.83	0.74	0.71	0.69	0.68	0.68	0.67	0.67
42	15.52	0.79	0.70	0.67	0.66	0.65	0.65	0.64	0.64
48	14.91	0.76	0.67	0.64	0.63	0.62	0.62	0.61	0.61

[a] $H = 8.5$; $B = 144.8$; $A = 116,825$; $HP = 4,500$.

will be noted that the (imaginary) envelope curve appears to have a minimum point, reflecting the onset of decreasing returns imposed by the fixed dimensions of the waterway. However, the envelope curve is quite flat in the neighborhood of its minimum point.

The question which naturally arises at this point is: If the depth and width of the waterway impose decreasing returns to scale on the tow beyond some critical tow size, what would be the effects on tow productivity of increasing these channel dimensions? These effects can be shown either in terms of the ton-mile per hour rate of output or in terms of costs per ton-mile. Table 8 illustrates the effects in terms of costs, a measure more directly related to the measurement of benefits from channel improvement.

It will be noted that the impacts of changes in channel width and depth are not independent. Roughly speaking, however, the effects of channel deepening are nearly exhausted at depths equal to four times the draft of the flotilla and the effects of channel widening at widths twice that of the flotilla.

AN EMPIRICAL PRODUCTION FUNCTION

A different approach to the estimation of the production function for the tow has been taken by Leininger [1963]. Using data from towboat log books, he estimated the functional relationship between the ton-mile per hour rate of output, TM,[7] and the horsepower and barge inputs, controlling for the following variables:

1. Flotilla average draft, H;
2. Flotilla configuration as measured by the length-to-breadth ratio, L/B;
3. Direction of travel ($W_2 = 1$ if upstream, $W_2 = 0$ if downstream);
4. Whether or not the towboat is equipped with the efficiency-increasing Kort nozzle ($W_1 = 1$ if Kort nozzle, $W_1 = 0$ otherwise).

[7] Ton-mile output on a particular trip was divided by operating hours, omitting time idle under repair or waiting orders.

The functional form used was

$$TM = \alpha_0 e^{\alpha_1 W_1} e^{\alpha_2 W_2} HP^{\alpha_3} H^{\alpha_4} N^{\alpha_5} \left(\frac{L}{B}\right)^{\alpha_6}. \tag{7}$$

This was fitted to a large number of observations on operations on the Ohio River, classified by season of operation and by river district.

The Leininger approach is not based on underlying technological relationships but is derived from operating data which reflect the results of many factors not taken into account by the preceding approach: the particular practices of pilots, the irregularities of the channel and current which can be turned to advantage, the effects of average weather conditions, the degree of congestion of the waterway and locks, and the impact on average rates of output of the required locking operations. For this reason, Leininger's functions should provide superior predictions of average rates of output for the waterway and conditions represented in his data. It becomes risky, however, to extrapolate beyond that range of circumstances.[8] It was found that river district had no significant impact on the parameters of (7) but that there were significant seasonal differences between the spring season when open-river conditions prevailed and the fall season when the river was in pool stage.[9] The functions estimated were as follows:

Spring season, $TM = 205 \; e^{0.11W_1} \; e^{-0.75W_2} \; HP^{0.27} \; H^{0.72} \; N^{0.84} \; (L/B)^{0.09}$;

Fall season, $TM = 280 \; e^{0.11W_1} \; e^{-0.15W_2} \; HP^{0.24} \; H^{0.84} \; N^{0.58} \; (L/B)^{0.23}$; \hfill (8)

where N is the number of barges. The first function was fitted to 121 observations and explained 82 per cent of the total variation in ton-miles per hour among the observations. The coefficient of variable W_1 was of the expected sign but significantly different from zero only at the 10 per cent level. The W_2 coefficient was highly significant, as were all others except that for (L/B). The second function was fitted to sixty observations and explained 85 per cent of the observed output rate variation. The W_1 and W_2 coefficients had the expected signs but neither was significant. The other coefficients were all highly significant. The differences between the two seasons are in keeping with prior reasoning. The first term indicates that, under all circumstances, the use of Kort nozzles increases overall output by 12 per cent ($e^{0.11} = 1.12$). The effect of direction of travel on output rates is much more dramatic under open-river conditions when stream currents are high: ceteris paribus, for upstream travel, a reduction from downstream output rates of 53 per cent in the spring ($e^{-0.75} = 0.47$) and 14 per cent in the fall ($e^{-0.15} = 0.86$). The impact of tow configuration appears to be

[8] No comparison of the predictive powers of the Howe and Leininger models was made because the Howe production function represents time under way at operating speed only, while Leininger's functions reflect locking operations also. For an account of the incorporation of the Howe function into a model which incorporates locking delays, see Chapter 5.

[9] On the Ohio River there are a large number of wicket dams—gratings which, when the flow is low, are raised to retard flow and increase pool depth. In the spring, when flows are high, the wickets are lowered and tows are no longer required to lock at those points but proceed directly down the open river.

less in the spring season than during the fall season, but the reasons for this are not clear.

Returns to scale ($\alpha_3 + \alpha_5$) to the horsepower and barge inputs appear to differ between the two river conditions: 1.11 for open river (spring season) and 0.82 for pool stage (fall season). Again, this is a reflection of the difference in channel width and depth, the effects of which have been noted earlier in this chapter, *plus* the impact of increased number of lockings and time spent in locking when the river is in pool stage. The latter factor was not included in the technological derivation of the tow's production function since the analysis was based solely on equilibrium speed under way. When locking is required, not only is more operating time spent locking or waiting to lock, but advantages from increasing the size of the tow are severely limited beyond the size that will pass through the lock chamber in one locking. When the tow exceeds this size, the flotilla must be broken into two or more sections and "double-locked," an operation requiring something in excess of twice the ordinary locking time. Furthermore, Corps of Engineers locking priority rules specify that any tow requiring more than a double locking must give way to other tows which may be waiting.[10] Thus, effective returns to scale (i.e., including the time spent locking) are severely reduced.

CONCLUSIONS

In this chapter, a technologically based production function was derived to assess the relationship between the barge tow and the environment in which it operates. This analysis was supplemented by the empirically derived production function estimated elsewhere by Leininger.

It was found that the use of the Kort nozzle on boats working the Ohio River results in an average increase in output rates (for otherwise similar equipment) of 12 per cent. As expected, the direction of travel has a definite impact on output rates, the impact being much more dramatic under open-river conditions than when the waterway is in pool stage. The average upstream-downstream difference reported by Leininger was 53 per cent under open-river conditions and 14 per cent in pool stage.

The effects of channel width and depth on the rate of output of the tow and on operating costs are dramatic when width and depth approach the breadth and draft of the barge flotilla. However, the favorable effects of increased channel width and depth appear to be largely exhausted at depths four times flotilla draft and at widths twice that of the flotilla.

The production function indicates that for larger flotillas there are substantial substitution possibilities between numbers of barges and towboat size. For smaller tows substitution is quite restricted and will, in practice, be highly

[10] Full regulations for locking procedures are found in U.S. Army, Corps of Engineers, *Regulations Prescribed by the Secretary of the Army for Ohio River, Mississippi River Above Cairo, Ill. and their Tributaries; Use, Administration and Navigation* (U.S. Government Printing Office, 1961).

constrained by navigational considerations, such as the need for extra power to maneuver or limits on flotilla length and width imposed by channel bottlenecks.

The production function and cost function analyses have shown that the tow is subject to increasing returns to scale up to a critical size which is a function of the channel width and depth. On very large rivers such as the lower Mississippi, increasing returns are probably experienced even beyond the maximum tow sizes currently in use (e.g., 9,000 horsepower and 40 barges), whereas on the smaller waterways the onset of decreasing returns is probably a very real operating constraint.

The cost function numerically derived from the production function reflected a decrease in average ton-mile costs up to 80- or 85-thousand ton-miles per hour. The long-run average cost curve was quite flat over the range of from 65- to 100-thousand ton-miles per hour. These figures reflect a particular waterway setting and flotilla configuration. If the waterway were larger, the minimum cost would be somewhat lower and would be attained at a higher rate of output.

3

RETURNS TO SCALE FOR THE FIRM

In the inland waterway industry, firms must manage the boats and barges they own or lease in such a way as to accommodate the demands of their customers. A firm may operate on several waterways of differing characteristics—i.e., channel dimensions, currents, numbers of locks, and degrees of congestion—although usually each firm is primarily engaged in barge transport on a particular river. If the origins, destinations, and timing of cargo permitted, the firm could presumably select a uniform tow size that would minimize the average costs of transport on its particular waterway, and this optimum size of tow would, from the observations of Chapter 2, become larger with increasing size and decreasing congestion of the waterway on which it operates.

Many operating tows fall below that optimum size because of the variety of origin-destination pairs and availability of cargo, although some tows may be made larger for scheduling reasons. This observation leads to the following links between returns to scale for the tow and overall returns to scale for the bargeline firm:

1. The tow must exhibit decreasing returns to scale beyond some size in a particular environment;

2. The larger and less constricted the waterway, the larger will be this critical tow size;

3. The larger the critical tow size, the larger will be the percentage of a firm's tows operating in the size range below critical size.

These statements lead to the following hypothesis:

> The larger and less constricted the waterway on which a firm carries out most of its operations, the greater will be the returns-to-scale parameter of that firm; i.e., the greater will be the increase in efficiency with increases in the size of the firm.

A natural test for this hypothesis would be to estimate production functions for several firms on different rivers to see whether or not the returns-to-scale parameters vary as indicated by the hypothesis.

The production function is a *technological* relationship between inputs and output, subsuming only the elimination of technologically inefficient combinations. This must be kept in mind when considering methods for measuring inputs, especially the capital inputs. A detailed consideration of this measurement problem and the related distinction between the firm's production function and its "planning" function is given in Appendix D to this chapter. We will only note here that in transport firms for which the capital stock consists of separable

34

units like towboats and barges, the rate of output at any instant of time is related only to that part of the capital stock actually in use. The idle portion of the capital stock is irrelevant to the production process, although it may be quite relevant to other aspects of the firm's planning; e.g., to meeting peak demands. Thus for the estimation of production functions, measures of the total *utilization* of the boat and barge stocks have been used as input measures.

Ultimately, however, the firm must make decisions regarding its stocks of equipment. In the longer term, too, the total output a firm is able to schedule, including peak demands, unanticipated demands, and unusual destinations, will depend upon the stocks of equipment on hand. Much more than technology is involved in the relationship between the stocks of capital inputs and output (say, on an annual basis). For example, the strategy the firm adopts for handling peak demands, the availability of equipment on short-term rental, and the time variability of demand will affect the stock-output relationship. Ultimately this capital stock-output relationship and its scale economies or diseconomies will determine the firm's ability to expand line-haul operations without incurring increasing inefficiency. It seems best to give such a relationship the name of "planning function" to distinguish it from the purely technological production function.

The distinction between a firm's production function and the planning function can be used to advantage in the following way: If both types of function can be estimated for firms typical of the industry, the sources of increasing or decreasing returns to scale can be isolated. If the production functions exhibit increasing returns to scale, but the planning functions exhibit decreasing returns, it seems reasonably clear that scheduling and other dynamic factors lie behind the overall decreasing returns of the firm. Should the opposite conditions prevail, it would indicate that the firm can schedule more efficiently and adapt to fluctuating demand more efficiently as the volume of output grows.

To carry out the steps outlined in the preceding paragraph, the following procedure was followed. (1) In order to estimate production functions using capital service inputs, monthly time-series data on three firms operating on different waterways were gathered; (2) to estimate the planning functions using capital stock inputs, combined cross-section time-series annual data on six firms were gathered. The latter type of data could be gathered because figures on stocks of towboats and barges are available in published form for all bargelines [U.S. Army, Corps of Engineers, *Transportation Lines . . .*], and most firms record ton-mile output, labor, fuel, and input costs.[1]

The input measures used referred only to the line-haul operations of the firms. Administrative labor and capital were omitted because firms engage to different degrees in various side activities—such as shipbuilding, boat and barge repair, warehousing, and the provision of dock facilities—and it was not possible to allocate the administrative inputs among transportation and the other activities.

[1] Several large firms whose inclusion in the cross-section had been anticipated did not, however, record ton-miles, but only tons of cargo carried.

The variables used in the analysis are as follows:[2]

> $TM \equiv$ cargo ton-miles generated (short tons);
> $B \equiv$ equivalent barge-days under way, full and empty;
> $H \equiv$ operating horsepower-hours;
> $L \equiv$ line-haul labor in man-hours, all grades;
> $F \equiv$ fuel oil in gallons;
> $O \equiv$ total boat operating hours;
> $C_B \equiv$ cost per equivalent barge day;
> $C_H \equiv$ total operating expense per operating horsepower hour.

The model used in the estimation process is given below.

$$TM = \alpha_0 \, B^{\alpha_1} \, H^{\alpha_2} \, u_1 . \tag{1}$$

$$B = \beta_0 \, TM^{\beta_1} \, B_{t-1}^{\beta_2} \, C_B^{\beta_3} \, u_2 . \tag{2}$$

$$H = \delta_0 \, TM^{\delta_1} \, H_{t-1}^{\delta_2} \, C_H^{\delta_3} \, u_3 . \tag{3}$$

$$F = \gamma_0 \, H^{\gamma_1} \, u_4 . \tag{4}$$

$$L = \epsilon_0 + \epsilon_1 \, O + \epsilon_2 \, H + u_5 . \tag{5}$$

Equations (1) to (3) constitute a complete model in the three endogenous variables TM, B, and H, with exogenous variables C_B and C_H, and predetermined variables B_{t-1} and H_{t-1}. Equations (4) and (5) are considered technological relationships relating the fuel and labor inputs to the boat input (operating horsepower hours). The details of the derivation of the model are presented in Appendix B to this chapter.

The parameters of the model (1) to (5) were estimated by using monthly data from three large common-carrier firms operating on three major legs of the Ohio-Mississippi system. The firms may be characterized as follows:

Firm	Type of operation	Number of observations
A	Common carrier, deep open waterway	35
B	Common carrier, moderately constricted waterway (in terms of width, depth, and number of locks)	28
C	Common carrier, highly constricted waterway	23

The model obviously calls for a simultaneous-equations estimating technique. Since the first three equations—the production function and the two factor demand equations—constitute a complete model in themselves, their parameters were estimated simultaneously by using the Generalized Classical Linear estimation procedure of Basmann.[3] The last two equations were individually estimated by standard least-squares techniques. The complete results are shown in

---

[2] Full definitions of these variables are given in Appendix A to this chapter.
[3] The General Electric 709 ECOMP Program, as adapted for the Purdue University IBM 7090, was used. The reference to that program is R. L. Basmann, *Guide for Use of 709 ECOMP Program*, RM 61 TMP–12 (Santa Barbara, California: General Electric Company, October 1961), Vol. 1.

Appendix C of this chapter; only the production functions and fuel and labor input equations are given here. Equations (1), (2), and (3) were fitted in logarithmic form, with lowercase letters designating logs to the base 10.

Firm A

$$tm = 2.19 + 1.16\,b + 0.30\,h, \qquad \hat{\rho} = 1.46 \tag{1a}$$
$$f = -0.44 + 1.14\,h \tag{4a}$$
$$L = 298 + 13.57\,O - 0.001\,H\,. \tag{5a}$$

Firm B

$$tm = 0.65 + 0.57\,b + 0.79\,h, \qquad \hat{\rho} = 1.36 \tag{1b}$$
$$f = -0.30 + 0.91\,h \tag{4b}$$
$$L = 21{,}371 + 8.13\,O - 0.001\,H\,. \tag{5b}$$

Firm C

$$tm = 2.06 + 0.46\,b + 0.64\,h, \qquad \hat{\rho} = 1.10 \tag{1c}$$
$$f = -0.13 + 0.85\,h \tag{4c}$$
$$L = 12{,}402 + 5.18\,O - 0.0002\,H\,. \tag{5c}$$

The coefficients of b and h give us the percentage by which output increases for a one per cent increase in each of these factors. The sum of the two coefficients indicates the percentage increase in output that would result if *both* inputs were increased by one per cent. This sum thus acts as a measure of the physical production economies or diseconomies which occur when the production process is expanded in scale.

The estimates of the returns-to-scale parameters, $\hat{\rho}$, for the three firms all exceed 1.0 and descend in the order indicated by the foregoing hypothesis: 1.46 for the deep water firm; 1.36 for the firm on the moderately constricted waterway; and 1.10 for the firm in the highly constricted waterway. Although the standard errors of the estimates are sufficiently large that none are significantly greater than 1.0, the results appear to be consistent with the hypothesis.

Labor is not freely substitutable for the boat (horsepower) or barge inputs, and fuel can be substituted for the boat input only to the very limited extent that boat engines can temporarily be overloaded. Thus the labor and fuel inputs did not appear as inputs in the production function, but are related directly to the measures of the boat (horsepower) input. For Firm A, which exhibits the highest returns-to-scale parameter, it will be noted that, when output is seasonally expanded, the increase in the rate of fuel input is more than proportional to the increase in horsepower operating hours, presumably because older, less efficient boats are utilized more intensively. The labor input, however, is nearly proportional to the number of boat operating hours.[4] Thus it becomes difficult to assess the overall returns to scale for Firm A in terms of all four inputs. The increases in labor and fuel inputs of firms B and C, however, are less than proportional to the boat inputs, reinforcing the increasing returns to scale found in the production function. It thus appears that all three firms exhibit increasing returns to scale in their line-haul operations.

[4] See Appendix B of this chapter for the interpretations of the effects of O and H on the labor input.

A second model was used to estimate what was earlier called the planning function of the bargeline's line-haul operations; i.e., the relationship between output and the capital stock inputs. Unfortunately, in the case of the three firms described above, it was not possible to gather sufficient data to continue the analysis.[5] Neither was it possible to obtain a sufficient number of observations to estimate separate functions for different waterway environments. It was finally decided to combine annual cross-section and time-series data for six large firms, including two of the three firms used in the preceding time-series analysis. This resulted in a total of forty-two observations over a period of thirteen years.
The variables used in the analysis are as follows:

$TM \equiv$ cargo ton-miles (annual total);
$B \equiv$ number of (195- by 35-foot) equivalent barges owned or leased by the firm;[6]
$HP \equiv$ total brake horsepower of all boats owned or leased by the firm;
$F \equiv$ total gallons of fuel;
$L \equiv$ total hours of line-haul labor;
$T \equiv$ time (1950 = 01, 1962 = 13);
$C_B \equiv$ annual cost per equivalent barge;
$C_H \equiv$ annual cost per horsepower (including depreciation and interest).

In the course of the year, the firm will normally lease equipment, especially barges. This is not a particularly troublesome point, for we are attempting here to measure the relationship between the long-run growth of output and corresponding growth of the stocks of boats and barges, as determined by technological considerations, factor prices, and the dynamic requirements of adapting to fluctuating demand. The firm's decision processes regarding optimal stocks of equipment undoubtedly take into consideration the possibility of leasing equipment from others for short periods.
The definitions of the variables used are fairly obvious, capital stocks, B and HP, replacing the capital flows used previously. C_B was computed as before, and C_H represents total annual towboat expense, including depreciation and interest, divided by total horsepower. The model used is as follows:[7]

$$TM = \alpha_0 \, B^{\alpha_1} \, HP^{\alpha_2} \, T^{\alpha_3} \, u_1 \,. \tag{6}$$

$$B = \beta_0 \, TM^{\beta_1} \, C_B^{\beta_2} \, T^{\beta_3} \, u_2 \,. \tag{7}$$

$$HP = \delta_0 \, TM^{\delta_1} \, C_{HP}^{\delta_2} \, T^{\delta_3} \, u_3. \tag{8}$$

$$F = \gamma_0 + \gamma_1 \, HP + \gamma_2 \, TM + \gamma_3 \log T + u_4 \,. \tag{9}$$

$$L = \epsilon_0 + \epsilon_1 \, HP + \epsilon_2 \, TM + \epsilon_3 \log T + u_5 \,. \tag{10}$$

---

[5] The capital stocks vary slowly over time, so that only annual data provide significant variations for analysis. The detailed records of each firm extended over too few years to permit the estimation of the planning function.

[6] Different classes of barges were converted to 195- by 35-foot equivalents by the ratio of their carrying capacity at a draft of 8.5 feet to that of a representative 195- by 35-foot barge.

[7] Again, equations (6) to (10) constitute a complete model in the endogenous variables TM, B, and HP and the exogenous variables C_{HP}, C_B, and T. Time has explicitly been introduced since the observations were taken over quite a long time span as well as over firms. During the observation period (1950–62), a significant amount of technological change took place in both boat and barge

In the results presented below, lowercase letters stand for logs (base 10) of variables. The standard errors are given below the estimates and the means of the variables are italicized below each equation. An asterisk by a coefficient indicates not significant.[8]

$$tm = .2924 + \underset{(.0842)}{.5282} \ hp + \underset{(.1179)}{.3396} \ b + \underset{(.0690)}{.1209} \ t \tag{6a}$$

2.1121   2.0099   1.9243   .8656

$$b = 1.7335 + \underset{(.0439)}{.7244} \ tm - \underset{(.0537)}{.3785} \ c_B + \underset{(.0476)}{.1681} \ t \tag{7a}$$

1.9243   2.1121   3.9226   .8656

$$hp = .9365 + \underset{(.0492)}{1.0891} \ tm - \underset{(.0890)}{.5983} \ c_{HP} - \underset{(.0558)}{.1640} \ t \tag{8a}$$

2.0099   2.1121   1.8136   .8656

$$F = 822,415 + \underset{(18.71)}{195.29} \ HP + \underset{(.1907)}{.6657} \ TM - \underset{(2.9905)}{5.5478} \ t \tag{9a}$$

4,133,257   13,918   1,611,787   .8656

$$L = 196,172 + \underset{(4.403)}{26.279} \ HP - \underset{(.0449)}{.0073^*} TM - \underset{(70,388)}{162,933} \ t \tag{10a}$$

409,078   13,918   1,611,787   .8656

---

design. The form of the planning function (6) used implies that this innovative activity has been neutral with regard to the two inputs.

Equations (6) to (8) are, for this model, approximations to the profit maximizing conditions for a firm selling its services in an imperfect market and buying its inputs in purely competitive markets. Utilizing the model of Appendix B to this chapter with appropriate change of variables, one gets the conditions:

$$B* = \hat{a}_0 \ \hat{a}_1 \ \beta_1 \ TM^{\hat{a}_1} \ C_B^{-1},$$

$$HP* = \hat{a}_0 \ \hat{a}_1 \ \beta_2 \ TM^{\hat{a}_1} \ C_{HP}^{-1},$$

$$TM = \alpha_0 \ B^{\alpha_1} \ HP^{\alpha_2} \ T^{\alpha_3}.$$

The lagged adjustment process postulated for the earlier model is not particularly appropriate when annual observations are used in lieu of monthly observations. While a large volume of business in month $(T - 1)$, accompanied by directional imbalance of traffic, is likely to leave a large legacy of empties to be redistributed in month T, it seems unlikely that this lagged adjustment will be important relative to the annual cargo flows. It also seems reasonable that optimal inputs $B*$ and $HP*$ should, factor prices remaining constant, decrease over time relative to TM output because of technological improvements. Thus we have informally amended the first two equations in this footnote to the forms given by (7) and (8). The presence of T in (7) and (8) provides a test of the hypothesis of the neutrality of technological improvement, too, for neutrality implies that β_3 and δ_3 should both be negative.

Equations (9) and (10) again are considered to be technological relationships between the fuel and labor inputs, the capital stock HP, and TM as a measure of the intensity of utilization of the capital stock. The inclusion of TM in those equations was felt to be necessary since the data period included years of cyclical prosperity and recession.

[8] Since several variables appear in the model in both raw and log form, they were scaled as follows:

$$HP \cdot 10^{-2}, \ F \cdot 10^{-5}, \ L \cdot 10^{-4}, \ TM \cdot 10^{-4}.$$

Equations (9a) and (10a) have been "unscaled." Means of log variables may differ by integer amounts from means of raw variables.

The planning function parameters in (6a) seem quite reasonable and are highly significant. The HP and B coefficients indicate ($\hat{\alpha}_1 + \hat{\alpha}_2 = .8678$) *decreasing* returns to scale in terms of the *stocks* of boats and barges maintained by the firms in our sample, although this sum is not significantly different from 1.0. It is clear that there has been a trend of increasing efficiency ($\hat{\alpha}_3 = .1209$), presumably because of technological improvements.[9] The hypothesis of neutral technological change embedded in the form of the planning function is, however, challenged by the difference in signs of t in (7a) and (8a). It appears that even after the effects of factor price changes have been taken into account, a tendency remains to increase the barge input relative to output and to decrease the horsepower input. This strongly suggests that technological improvements have been "horsepower saving"; i.e., biased against the boat input.

The fuel and labor functions have significant time trends (log T appears in each equation), reflecting the general technological improvement. The price of fuel (per gallon) has remained steady or has fallen somewhat over the period 1950–62. The average wage of line-haul labor, however, has risen to approximately 250 per cent of its 1950 level. This wage increase has resulted in substitution against the HP input. The time trends in equations (7a) and (8a) also hint that technological change which is "horsepower saving"—i.e., biased against the boat input—may have been induced by this rapid labor cost increase.

These results indicate that when the full stocks of equipment held against cyclical and random variations in demand are taken into account, the related returns-to-scale parameter of the firm is less than that of the production function itself. The results unfortunately do not provide information on how this parameter may vary in different waterway environments.

A final analysis of returns to scale, taking into account all inputs of the firm related to water transportation including personnel and equipment involved in scheduling, administration, etc., has been carried out in terms of the costs of fifteen bargelines in each of the years 1957 and 1962.[10] The variations in cost relevant here are those longer-term changes which take place as the firm grows or adapts to permanently higher rates of output.

Costs can vary with output for two reasons: (1) the production and administrative processes may become more efficient in the physical sense of more output per unit of input; (2) the prices at which inputs can be purchased may vary with the quantities purchased. While the second type of cost variation is of direct interest to the firm and may constitute a definite motivation for expansion, it may reflect only the superior bargaining power of larger firms rather than any reduction in the use of resources. For this reason, most cost studies attempt to eliminate this source of cost variation by repricing or deflating certain com-

[9] "Presumably" because it is also possible that the apparent increase in efficiency is due to a changing commodity mix.

[10] The basic source for data for the calendar years 1957 and 1962 was *Transport Statistics in the United States, Part 5: Carriers by Water* (U.S. Interstate Commerce Commission, Bureau of Transport Economics and Statistics, 1958 and 1963). The firms utilized in the analysis were all Class A carriers; i.e., those having annual operating revenues which exceed $500,000.

ponents of cost. A better way is simply to allow prices to appear as independent variables in the cost function along with the rate of output.

The data used in this study did not permit any adjustments to be made to the costs of the different sized firms, so that the observed cost variations may be caused both by changes in physical efficiencies and by variations in prices paid for inputs. Separate cost functions were fitted to the data of those two years to test for the stability of the parameters.

The definition of cost used here is the same as that used by the Interstate Commerce Commission: total waterline operating expense,[11] including maintenance expenses; depreciation and amortization of transportation property; line service expenses (wages, fuel, food, etc.); terminal service expenses; traffic expenses (scheduling and advertising); general expenses (administrative overhead); casualties and insurance; operating rents (charters of transport equipment and property); payroll, income, and property taxes. Letting C_{it} represent this expense item for the ith firm in year t and Z_{it} represent the total tonnage carried (the only available output measure), the following equations were fitted:

$$C_{it} = a_0 + a_1 Z_{it} + e_{it} \tag{11}$$

$$C_{it} = b_0 Z_{it}^{b_1} u_{it} ; \tag{12}$$

where e_{it} and u_{it} represent the appropriate forms of residual random variation. The following results were obtained:

$$C_{i,57} = 25{,}448 + \underset{(0.22)}{1.01} Z_{i,57} \qquad R^2 = 0.62 . \tag{11a}$$

$$C_{i,62} = 30{,}274 + \underset{(0.20)}{1.05} Z_{i,62} \qquad R^2 = 0.67 . \tag{11b}$$

$$C_{i,57} = 30.25 Z_{t,57}^{0.73} \qquad R^2 = 0.65 . \tag{12a}$$
$$\underset{(0.15)}{}$$

$$C_{i,62} = 28.04 Z_{t,62}^{0.75} \qquad R^2 = 0.76 . \tag{12b}$$
$$\underset{(0.12)}{}$$

For both forms of the cost function, the parameter values are about the same in each year. The goodness of fit is somewhat better for the multiplicative form (12a and 12b). Both forms indicate a declining cost per ton as the firm grows larger, although the constant terms in (11a) and (11b) are small relative to the range of values of C_i (approximately \$1,200,000 to \$18,000,000) and Z_i (approximately 120,000 tons to 14,500,000 tons). A ton-mile measure of output would have been more appropriate but was not available from published sources.

Conclusion. The basic hypothesis of this chapter was that, because of the relationship between the waterway environment and the efficiency of the tow, the larger and less constricted a waterway, the greater would be the returns to

---

[11] The cost figures are taken from Table 1 of each issue of *Transport Statistics*. . . . The precise definitions of each category of operating expense can be found in U.S. Interstate Commerce Commission, "Annual Report Form K–A, Inland and Coastal Waterways (Class A and Class B carriers)," available from the U.S. Government Printing Office.

scale exhibited by the firm. To test this, data on bargeline firms were used to estimate production functions for the line-haul operations of the firms, planning functions (using line-haul capital stock inputs rather than measures of capital goods services), and total cost functions.

The production function analysis was consistent with the hypothesis stated above. Returns to scale in line-haul operations were 1.46 for the firm operating on a large, open river, 1.36 for the firm on a moderately constricted river, and 1.10 for a firm operating on a severely restricted waterway.

The planning function analysis indicated, however, that there is a decline in efficiency of the utilization of the stocks of boats and barges as firm size increases. The scale parameter for the planning function was 0.87. It was also seen that there had been a time trend of increasing efficiency which has undoubtedly acted to offset the decreasing returns to some extent. These results still refer only to the line-haul operations of the bargeline firm.

A cost function approach was taken to estimating returns to scale for the overall operations of the firm, including its administrative and overhead activities. The results showed increasing returns to scale overall, with very little difference in overall efficiency over time.

Thus it appears that with volumes of traffic and waterway conditions as they have been to date, firms have been able to increase their efficiency as volume has increased. The production function results show that firms can increase their technical efficiency as the size of their operations increases and by being provided with a deeper and less congested waterway. This seems to be offset to some extent by increased scheduling difficulties or other administrative complications which result in a tendency of stocks of equipment to grow more than in proportion to the growth in output. These difficulties are, in turn, offset by the ability to spread overhead costs and perhaps by more advantageous purchasing of inputs, so that total unit costs fall with increasing firm size.

To what extent these overall scale economies will be reflected in the performance of private bargeline operating costs in the future depends largely upon the degree of congestion which firms will find on the waterways. This issue and the matter of public investment in waterway improvements are considered in Chapter 5.

APPENDIXES TO CHAPTER 3

A. DEFINITIONS OF VARIABLES USED IN MODELS
OF THE PRODUCTION FUNCTION OF THE FIRM

Ton-miles is perhaps an inadequate measure of output since a ton-mile figure may represent different ton and mile combinations, each of which may require different inputs. A further difficulty is that different commodities are frequently shipped in different lot sizes. For example, grain is always shipped in full barge loads, but high-value steel products may be shipped in partial barge loads because of the inventory decisions of the shippers. Since both require a barge and because a partially loaded barge will require nearly as much motive power as a full barge, the input-output relationships of operations specializing in the two commodities would be quite different. This is but one example of ton-mile heterogeneity. Clearly, a ton-mile upstream differs from a ton-mile downstream, as does a ton-mile in shallow water from a ton-mile in deep water. Thus it is necessary that the ton-mile "mix" be fairly constant over time if we are to obtain comparable measures of output.

The barge-day input was computed by reducing all types of barges to their equivalent in terms of the cargo capacity of a 195- by 35-foot barge at a draft of 8.5 feet. Thus a "standard" 175- by 26-foot barge which carries 900 tons at 8.5-foot draft is equivalent to (900/1,350) barges of the 195- by 35-foot class. The number of under-way days for each class of barge was then multiplied by the equivalence factor for that class, and the resulting figures summed over all barge classes. Up to fifteen different types of barges may be used by a firm, although the 195- by 35-foot and 175- by 26-foot classes predominate.

Operating horsepower-hours represents the total number of hours that boats were engaged in operations, including locking time and waiting to lock, each boat weighted by its rated brake horsepower. Time spent in overhaul or idle waiting orders was excluded. Boat operating hours, O, is simply the *unweighted* total of operating hours. Labor is measured in man-hours and represents time on the boat. On all but local harbor work, at least two crews are on board. No distinction was made among grades of labor.

Fuel is measured in gallons of No. 2 diesel fuel equivalent. Some boats mix different types of fuel, so a reduction to a standard grade on the basis of Btu equivalents was necessary.

The variables C_B and C_H represent our approximations to factor prices. In the short run analyzed here, the firms had nearly constant stocks of barges and boats. Thus, as they considered whether or not to accept business, they had to consider only the variable costs involved, including any opportunity costs created by the possibility of very short-term leasing of equipment to other firms. The variable barge costs are very low but there is usually a ready per diem market in which idle barges can be loaned at a negotiated per diem rate. From barge utilization reports and expense statements, we constructed a series of average per diem rates per equivalent barge day. The C_H variable represents total variable operating expense per horsepower hour and includes all expenses related to crew, fuel, and maintenance. It does not include depreciation or return on investment. Since the towboat input was measured in horsepower hours, the input "price" was stated in the same units. While towboats may be leased to others, only long-term (one year or more) leases are frequent.

Two additional comments about the nature of the bargeline business are in order before we describe the model used and the empirical results obtained. A common carrier is permitted to operate only on specified reaches of the river and canal system. Much of the cargo a common carrier transports, however, may be destined for points located on waterways that its own boats do not serve. Thus they have "interchange" and "pool" agreements whereby their boats carry other firms' barges and others carry their barges. Further, as mentioned earlier, barges are frequently loaned to or from other lines on a per diem basis. It is possible that cargo solicited by Firm A is carried in barges belonging to Firm B and towed by a boat of Firm C. Such a situation would be exceptional, but serves to illustrate the possibilities. To make our measures of output and inputs as consistent as possible, the TM measure refers to total cargo ton-miles produced by the firm's own boats only, whatever the origin of the cargo; i.e., regardless of whether the carrying was being done for the carrier's own account or for other bargelines. The horsepower-hour input refers only to the firm's own boats, and the equivalent barge-day input refers to total usage of the firm's own (owned) barges, plus those leased and used on a per diem basis.

B. DERIVATION OF THE EQUATIONS OF THE
COMPLETE PRODUCTION FUNCTION MODEL

The basis for the first three equations follows: bargelines operate in an imperfect product or service market wherein the demand for the firm's services is quite responsive to the price charged. This demand elasticity stems both from bargeline customers' price consciousness and fickleness and from the often ready alternative of shipping by rail. Thus, it seems reasonable to approximate the firm's demand curve by

$$TM_D = a_0 P^{a_1}, \ a_1 < 0 \ . \tag{B1}$$

Total revenue associated with (B1) is then

$$TR(TM) = \left(\frac{1}{a_0}\right) \frac{1}{a_1} \ TM^{1+\frac{1}{a_1}} \equiv \hat{a}_0 \ TM^{\hat{a}_1} \ . \tag{B2}$$

The firms of the industry appear to buy their labor and fuel inputs under purely competitive conditions. The opportunity cost for barges is set (as a per diem rental rate) under conditions which are largely beyond the control of the individual firm. Thus, we characterize the short-run behavior of the firm as attempting to maximize

$$\pi\,(TM, B, H) \equiv \hat{a}_0 \ TM^{\hat{a}_1} - C_B B - C_H H$$

$$\text{subject to } TM = \alpha_0 \ B^{\alpha_1} H^{\alpha_2} \ . \tag{B3}$$

Forming the Lagrangean function related to (B3), differentiating partially with respect to TM, B, H, and λ, and eliminating λ from the equations, one gets:

$$B^* = \hat{a}_0 \ \hat{a}_1 \ \alpha_1 \ TM^{\hat{a}_1} \ C_B^{-1}$$

$$H^* = \hat{a}_0 \ \hat{a}_1 \ \alpha_2 \ TM^{\hat{a}_1} \ C_H^{-1}$$

$$TM = \alpha_0 \ B^{\alpha_1} \ H^{\alpha_2} \ . \tag{B4}$$

The first two equations of (B4) are the usual first-order maximization conditions, for a reshuffling of terms yields the usual marginal revenue product equals factor price for each of the two inputs. It is reasonable to assume, however, that the precise equilibrium values of (B4) will not be attained, both because we have omitted stochastic elements from our model and (most important) because the preceding period (month) always leaves a legacy of a geographical distribution of equipment stemming from traffic imbalance which must be redressed during the current period. The following adjustment lag seems appropriate:[12]

[12] Taken from Ta-Chung Liu and George Hildebrand, "Manufacturing Production Functions: A Review and Another Attempt at Estimation" (Paper given before the Econometric Society, Pittsburgh, December, 1962).

$$\frac{B_t}{B_{t-1}} = \left(\frac{B_t^*}{B_{t-1}}\right)^{j_B} ; \frac{H_t}{H_{t-1}} = \left(\frac{H_t^*}{H_{t-1}}\right)^{j_H}.$$ (B5)

$$0 < j_B \leq 1 \qquad\qquad 0 < j_H \leq 1$$

Substituting from (B4) into (B5) and solving for B_t and H_t yields:

$$B_t = (\hat{a}_0\, \hat{a}_1\, \alpha_1)^{j_B}\, TM^{\hat{a}_1 j_B}\, B_{t-1}^{1-j_B}\, C_B^{-j_B}$$

$$H_t = (\hat{a}_0\, \hat{a}_1\, \alpha_2)^{j_H}\, TM^{\hat{a}_1 j_H}\, H_{t-1}^{1-j_H}\, C_H^{-j_H}.$$ (B6)

Adding the random variables u_2 and u_3 to (B6) yields, in greater detail, the equations (2) and (3). The detail of (B6) exhibits various restrictions on the parameters which are implied by the model.

The justifications of (4) and (5) are largely technical. The fuel consumption curve typical of marine diesel engines (lbs fuel/brake horsepower *versus* engine brake horsepower) is fairly flat over the design range.[13] Thus the fuel consumed by a single engine is approximately proportional to brake horsepower-hours. If a firm had a fleet of identical boats, one could expect fuel consumption to be approximately proportional to total brake horsepower-hours. If, however, a firm keeps some older boats in standby status for peak periods or pulls small boats out of harbor service for line service at such times, one might anticipate an increase in fuel consumption that was more than proportional to brake horsepower-hours. Equation (4) permits a test of this hypothesis.

The labor input equation, (5), reflects our attempt to characterize the composition of the horsepower-hour input. For a given value of O (total operating hours, all boats), the H variable (total horsepower-weighted operating hours, all boats) measures the distribution of operating time between smaller and larger boats. For example:

$$O = 100\begin{Bmatrix}50 \text{ hours for 1,000-hp boats}\\50 \text{ hours for 5,000-hp boats}\end{Bmatrix} H = 300,000$$

$$O = 100\begin{Bmatrix}25 \text{ hours for 1,000-hp boats}\\75 \text{ hours for 5,000-hp boats}\end{Bmatrix} H = 400,000.$$

For given O, a larger (smaller) value of H indicates greater (less) relative use of larger boats. The different types of service in which the different sizes of boats are used dictates different crew sizes. Harbor service (flotilla collection and make-up) and line service in areas which require frequent pick-ups, drops, and locking dictate larger crews, particularly deck force. Since smaller boats are used for these types of service, it is not unusual to observe a 5,000- or 6,000-hp boat carrying the same sized crew or even a smaller crew than, say, an 1,800- or 2,500-hp boat. For this reason, it seems reasonable to expect that, for given O, larger H will lead to a smaller labor input.

C. ESTIMATES OF PARAMETERS OF PRODUCTION
FUNCTION MODEL FOR THREE FIRMS

The equations below are presented in the forms in which they were estimated, lowercase letters designating logs to the base 10. The standard errors of the estimates of the parameters are given below in parentheses, and the sample means of the variables are listed in italics below each variable in each equation.

---

[13] See, for example, General Motors, *Catalog G–140D* (Electro-Motive Division, General Motors Corporation, LaGrange, Illinois, no date).

Firm A

Common Carrier, Deep Open Waterway

$$-1tm_t + 1.16149b_t + 0.30128h_t + 2.19130 \; ; \tag{1}$$
$$(0.29397) \quad (0.21396) \quad (1.17611)$$

8.3909   3.4566   2.2496

$$0.09808tm_t - 1b_t - 0.05624c_B + 0.51590b_{t-1} + 0.90492 \; ; \tag{2}$$
$$(0.35394) \qquad\qquad (0.12712) \quad (0.35202) \quad (1.98473)$$

8.3909   3.4566   0.984   3.4596

$$0.67448tm_t - 1h_t - 0.73377c_H - 0.02068h_{t-1} + 2.69387 \; ; \tag{3}$$
$$(0.15573) \qquad\qquad (0.12716) \quad (0.14552) \quad (0.68832)$$

8.3909   2.2496   1.3065   2.2539

$$L = 298 + 13.5659\,O - 0.00124H \qquad\qquad R^2 = 0.93 \; ; \tag{4}$$
$$(0.9003) \qquad (0.00034)$$

71,059   6,876   18.2 × 10⁶

$$f = -0.4394 + 1.1361\,h \qquad\qquad R^2 = 0.60 \,. \tag{5}$$
$$(0.1603)$$

2.1163   2.2496

Firm B

Common Carrier, Moderately Constricted Waterway

$$-1tm_t + 0.57120b_t + 0.78669h_t + 0.65430 \; ; \tag{1}$$
$$(0.81053) \quad (0.77987) \quad (3.00155)$$

8.3289   3.3865   2.2770

$$0.63201tm_t - 1b_t - 0.00916c_B + 0.11484b_{t-1} - 2.25106 \; ; \tag{2}$$
$$(0.13876) \qquad\qquad (0.12138) \quad (0.14024) \quad (0.88104)$$

8.3289   3.3865   1.0162   3.3864

$$0.52767tm_t - 1h_t - 0.36649c_H + 0\;.09911h_{t-1} + 2.67992 \; ; \tag{3}$$
$$(1.45136) \qquad\qquad (1.59143) \quad (1.06233) \quad (6.60462)$$

8.3289   2.2770   1.2783   2.2737

$$f = -0.3005 + 0.9051\,h \qquad\qquad R^2 = 0.68 \; ; \tag{4}$$
$$(0.1213)$$

1.7605   2.2770

$$L = 21,371 + 8.1290\,O - 0.00118\,H \,. \qquad R^2 = 0.90 \,. \tag{5}$$
$$(1.8883) \qquad (0.00067)$$

55,375   6,969   19.1 × 10⁶

Firm C

Common Carrier, Highly Constricted Waterway

$$-1tm_t + 0.46273b_t + 0.63705h_t + 2.06406 \; ; \tag{1}$$
$$(0.50195) \quad (0.57395) \quad (2.41798)$$

7.9872   3.2515   1.9071

$$1.02388tm_t - 1b_t - 0.20112c_B + 0.11827b_{t-1} - 5.06311 ; \quad (2)$$
$$(0.13083) \quad\quad\quad (0.17674) \quad (0.09984) \quad (1.08262)$$

7.9872   3.2515   1.1909   3.2514

$$0.72566tm_t - 1h_t - 0.23878c_H + 0.17957h_{t-1} + 0.17206 ; \quad (3)$$
$$(1.21883) \quad\quad\quad (2.31711) \quad (0.19467) \quad (12.9987)$$

7.9782   1.9071   1.4714   1.9068

$$f = -0.1279 + 0.8479h \qquad\qquad R^2 = 0.82 ; \quad (4)$$
$$(0.0881)$$

1.4891   1.9071

$$L = 12,402 + 5.1776 O - 0.00023 H \qquad R^2 = 0.88 . \quad (5)$$
$$(1.8426) \quad\quad (0.00118)$$

40,484   5,785   8.2 × 10⁶

The identifiability test statistics for the three equations of the simultaneous part of the model are given in Table 9. The identifiability test statistic for the production function of Firm A was such (3.90) that the null hypothesis of identifiability is rejected at the t per cent level ($F_{2, 30, .95} = 3.32$), although it would not be rejected at 1 per cent. The identifiability hypothesis is not rejected for the other production functions. However, the horsepower demand functions appear not to be identified for firms B and C.

TABLE 9.   Identifiability Test Statistics and 5 Per Cent Critical Levels

Firm and equation	Test statistics	5 per cent critical level	Degrees of freedom
Firm A: (1)	3.90	3.32	2,30
(2)	0.78	4.17	1,30
(3)	0.59	4.17	1,30
Firm B: (1)	3.24	3.42	2,23
(2)	1.05	4.28	1,23
(3)	198.32	4.28	1,23
Firm C: (1)	1.66	3.55	2,18
(2)	0.83	4.41	1,18
(3)	46.16	4.41	1,18

D. THE MEASUREMENT OF CAPITAL INPUTS

The classic problem of measuring capital usually arises because of the heterogeneity of the goods under consideration both with respect to types of goods and to age. In production studies, a problem remains even when the capital goods are homogeneous so that an unambiguous measure of capital can be defined. This problem is the determination of whether the *stock* of capital goods or a measure of the *flow* of the services of the capital goods should be introduced as an argument in the production function.

Three cases come to mind in which the resolution of this capital input problem might be handled differently:

1. *The Case of Proportionality.* Production takes place either continuously over time or is carried on during the same number of hours per "week," the entire capital stock is utilized, and the intensity of utilization is always uniform. In this case, any measure of

capital services would be proportional to the capital stock and it would make no essential difference whether capital services or capital stock were used as the capital input in the production function.

2. *The Case of Integrated Capital Goods.* The capital stock and production process are of such a nature that whenever any part of the capital stock is in use, all of it must be in use, but the intensity of utilization can be varied. For example, when gas is transmitted by pipeline, pipelines and pumping stations must be used simultaneously, but varying amounts of product may be transmitted by varying the pumping speed. Another example is steam-electric generation. Whenever electricity is being produced by a single boiler-turbine-generator combination, all of the capital goods units are in simultaneous operation, but the level of output can be varied by changing the flow of steam from boiler to turbine. In these cases, the entire capital stock affects the way in which the variable inputs are combined at all intensities of utilization.

3. *The Case of Separable Capital Goods.* The capital stock consists of many relatively small homogeneous units which, when utilized at all, are utilized at a uniform intensity so that as the level of production varies, the number of units of capital goods being utilized also varies. The capital goods not currently in use in no way affect the actual production process. (They may be held against peak requirements or uncertain demand.) For example, in railroad transportation, two types of capital equipment are used; locomotives and cars. During any period of observation, the actual output of ton-miles is related to the actual utilization of locomotives and cars, but is not related to that part of the capital goods stocks which lay idle during the period.

The theoretical refinement made possible by bringing the capital stock directly in as an argument of the production function is seen in Smith's work on production planning and optimal investment [Smith, 1961]. The optimal investment program can be deduced directly from cost minimization subject to the production function, both for the static and certain dynamic cases. Smith has argued that in many cases the rate of flow of the variable inputs, such as labor, energy, etc., implies a rate of utilization of the capital stock, so that the capital stock may legitimately be introduced into the production function. This seems reasonable in what we have called the case of integrated capital goods. In the case of gas pipelines, an increase in the energy flow to the pumps clearly implies more intensive utilization of the stock of capital goods, and what one means by more intensive utilization is unambiguous. The argument seems much less applicable to separable capital goods.

In the case of separable capital goods (as in the transportation industries), the rate of output at any *instant* of time is directly related to that part of the capital stock currently in use. The idle part of the capital stock is completely irrelevant to the actual *production* process (though it may be quite relevant to other aspects of the firm's planning such as meeting peak loads and improving the firm's competitive position vis-à-vis other carriers). Given the capital goods in use, changing the size of the total capital stock would in no way affect the rate of output. Further, we never observe a production process at a point in time, but only the cumulative results over a finite time interval. During any such interval, variations in output rate occur with some corresponding variations in the utilization of capital. It would not be adequate to measure capital utilization simply by counting the number of units that were utilized at some time during the period, for some may have been used continuously while others may have been used for lesser amounts of time. For these reasons, it seems much more appropriate when estimating production functions for the case of separable capital goods to measure capital inputs as flows of services rather than as stocks; e.g., machine-hours instead of machines, locomotive-hours or horsepower-hours instead of locomotives or their deflected dollar value, etc.

The preceding argument seems valid when we stick to the concept of a production function as a technological relationship between inputs and output. However, firms do operate under conditions of fluctuating and uncertain demand for their services and they do, therefore, face decisions regarding the optimal *stocks* of capital equipment to be kept on hand. It appears entirely possible that economies or diseconomies of scale may be present in the production function itself (with capital services as inputs), while additional but quite independent economies or diseconomies may accrue to the firm in terms of the size of the capital stock which must be held in order to adapt optimally to a fluctuating and uncertain environment. Note, however, that the decisions regarding the capital stock, while involving the production function, incorporate much more than just this technological relationship: they involve all of the variables which must be considered in dynamic long-run profit maximization.

To provide an example in barge transportation, it is possible that a firm experiences increasing returns to scale in the production process itself (because a larger volume of cargo permits a higher proportion of larger, more efficient tows) but experiences increasing complexities in scheduling its equipment, so that the *stock* of equipment must be increased more than in proportion to the increase in the rate of output.

In order to understand the structure of the industry (i.e., the size distribution of firms) or the effects of mergers on efficiency, it seems relevant to study the relationship between the rate of output and inputs using capital stocks. However, this is not a production function in the usual sense and should perhaps be given some other name (e.g., planning function).

4

DETERMINATION OF EQUIPMENT REQUIREMENTS FOR THE BARGELINE: ANALYSIS AND COMPUTER SIMULATION

How much equipment (boats and barges) is required in order to carry out a particular transport task with a high probability of success? In considering this question in this chapter, the transport task is defined in terms of the numbers of bargeloads of cargo arising at each of n ports and requiring transportation to specified destinations. The loads of cargo are specified not as fixed numbers per unit time but as random variables with time-invariant means. The times required for loading and unloading barges are also treated as random variables. The objective of the analysis is to find a practical way of determining the minimum numbers of barges and boats consistent with *statistical equilibrium* of the system; i.e., a condition in which no ever-growing queues of cargo are present. As expected, the analysis shows that substitution possibilities exist between boats and barges, much as was demonstrated earlier for the individual tow. This, then, means that for given equipment costs, a minimum-cost fleet composition may be sought.

The method, which makes some simplifying assumptions to facilitate the analysis of this complex problem, is checked against the computer simulation of a realistic, randomly variable problem in the second section of this chapter.

In contrast to the problems in the literature on equipment scheduling where fleet size is assumed given, the present study makes assumptions about the scheduling rules used by the bargeline and centers attention on the compatibility of a particular fleet with a (probabilistically) specified transport task.

APPROXIMATIONS TO EFFICIENT FLEETS

It is assumed for notational convenience only that the segment of the river over which the bargeline operates has no branches. There are P linearly ordered ports served; they are sequentially labeled from 1 through P. The basic time unit is a day. A month is defined to be a period of exactly thirty days.

The amount of cargo made available to the bargeline per day for transit between each pair of ports is regarded as a random variable with a mean that does not change during the period considered. The expected monthly arrival rate, in bargeloads, of cargo originating at port i to be transported and delivered to port j is known and is represented by the symbol q_{ij}, for $i, j = 1, \ldots, P$. (For simplicity, define $q_{ii} = 0$ for $i = 1, \ldots, P$.) The expected daily arrival rate of cargo at port i for port j is $q_{ij}/30$ for $i, j = 1, \ldots, P$ and is assumed to be time-invariant for the duration considered. The matrix (q_{ij}) is called the expected monthly load pattern.

The numbers of days required for loading and for unloading are also as-sumed to be random variables. The *expected* loading and unloading times are known and depend only on the particular port at which the loading and unload-ing occur; they are denoted d'_k and d''_k respectively for port k $(k = 1, \ldots, P)$. Facilities for loading and unloading are assumed to be unlimited so that a barge may be unloaded as soon as it arrives at its destination. Similarly, a barge can begin loading as soon as it and the cargo it is to carry are both available.[1]

The number of days required for travel from a given port to a specific adjacent port is assumed to be a known constant. Let d_{kh} denote the transit time, in days, over the directed leg from port k to adjacent port h, for $k = 1, \ldots, P$ and $h = k - 1, k + 1$. (Of course, h takes on the values $1, \ldots, P$ only. In the sequel it will be assumed that the reader will supply such necessary modifications on the ranges of indices.)

All barges are perfect substitutes for one another and all boats are identical. There is an upper bound, denoted c, on the number of barges which may travel together in one tow.

The given data are, in summary: the linearly ordered ports $1, \ldots, P$; the expected load pattern q_{ij}, $i, j = 1, \ldots, P$; the mean loading and unloading times d'_k, d''_k, $k = 1, \ldots, P$; the constant transit times d_{kh}, $h = k - 1, k + 1$, $k = 1, \ldots, P$; the maximum flotilla size c.

On any given day, the state of the line-haul system with a specified fleet can be described jointly by the following quantities:

- the location of each of the boats;
- the destination and status (empty or loaded) of each barge in each tow;
- the number of loaded barges originating at port i and destined for port j which are at port k waiting for movement, for all ports i, j, and k;
- the number of barges awaiting unloading and being unloaded at each port;
- the number of empty barges at each port;
- the number of loads of cargo at port i destined for port j which are awaiting loading, for all ports i and j;
- the number of barges being loaded at each port.

The system will be said to be in a *statistical equilibrium* if the expected values of the components describing the state of the system are periodic func-tions of time; i.e., either constant or regularly recurring values. The equilibrium reached may depend upon the initial state of the system. The system, consisting of stocks of equipment and specified scheduling rules, is said to be *feasible* if the sum of the expected rates of cargo delivery to the destinations equals the sum of the expected rates at which loads arrive into the system. If the system attains a statistical equilibrium (i.e., if all characteristics of the system have finite ex-

---

[1] In practice barges are not always unloaded immediately. However, for the purposes of the present analysis it is not necessary or useful to distinguish between time spent awaiting unloading and time spent in the unloading process. It will simplify the exposition and the computer simulation to combine these two times and speak as though this total random time is actually spent in unload-ing. A similar expositional shortcut is made for loading.

pected values), it will be feasible. To see that this is so, consider what can happen to a shipment after it has been received by the bargeline: it must be awaiting loading, being loaded, loaded, awaiting movement, in tow, awaiting unloading, unloading, or else it has been delivered. Since in an equilibrium, by definition, the expected number of loads in each of these categories except the last cannot grow, the rate of flow of cargo out of the system must be equal to the rate of flow of cargo into it. That is, the system is feasible. The discussion which follows applies to a system only after a statistical equilibrium has been reached.

Reference will be made to ports of origin (PO), ports of destination (PD), and intermediate ports. Every port can be a port of origin and of destination. For a loaded barge, the PO and PD are those of the cargo it carries. An intermediate port is a port located between the PO and the PD of the barge or shipment under discussion.

Empty Barge Movements Induced by the Load Pattern

By assumption, there are no circuits in the transportation network and there are only finite stocks of boats and barges available. Therefore the average monthly flow of both barges and boats over a leg of the network must be the same in each direction if the system is in a statistical equilibrium. That is, the number of loads which must travel from port k to port $k + 1$ on their way from the port of origin to port of destination *plus* the number of empty barges going from port k to port $k + 1$ must equal, on the average, the number of loads plus the number of empties going from port $k + 1$ to port k, for any port $k = 1$, ..., $P - 1$. This condition may be stated algebraically as follows:

$$\sum_{i=1}^{k} \sum_{j=k+1}^{P} q_{ij} + M_{k,k+1} = \sum_{j=1}^{k} \sum_{i=k+1}^{P} q_{ij} + M_{k+1,k} \text{ for } k = 1, \ldots, P - 1, \qquad (1)$$

where M_{kh} denotes the expected number of empty barges crossing the directed leg from port k to port h per month. In equations (1) it is assumed that the loaded barges crossing the leg from port k to port $k + 1$ contain loads originating at one of the ports $1, \ldots, k$ and terminating at one of the ports $k + 1, \ldots$, P. The expected number of loads crossing the leg from port $k + 1$ to port k is found analogously.

To see the necessity of equations (1) for a statistical equilibrium, let

$$R_k = \left[\sum_{i=1}^{k} \sum_{j=k+1}^{P} q_{ij} + M_{k,k+1} \right] - \left[\sum_{j=1}^{k} \sum_{i=k+1}^{P} q_{ij} + M_{k+1,k} \right] \text{ for } k = 1, \ldots, P - 1,$$

and suppose first that R_k is positive for some k. Then barges will collect at one or more of the ports $k + 1, \ldots, P$ at an expected rate of R_k per month. Suppose that there are b barges in the fleet. After an interval of expected length b/R_k months, all b barges will be distributed in some fashion among the ports $k + 1$, ..., P. Then it will no longer be possible to maintain a rate of flow over the leg from port k to port $k + 1$ which exceeds the rate of flow over the leg from port

$k + 1$ to port k. That is, there is a contradiction to the equilibrium condition supposed. Where R_k is negative for some value of k, the demonstration of a contradiction is similar, with barges collecting at ports $1, \ldots, k$. Hence it is seen that equations (1) must hold in a statistical equilibrium.

Since it is assumed that the distributions of cargo interarrival times at ports of origin are time-invariant, there need be no cross-shipment of empty barges. That is, flows of empty barges may be chosen such that

$$M_{k,k+1}M_{k+1,k} = 0, \text{ for } k = 1, \ldots, P - 1. \tag{2}$$

It is possible to imagine particular events which might make it advisable, in some sense, to deviate temporarily from the policy represented by equations (2). Nevertheless equations (2) will be taken as representative of the situations to be considered.

Equations (1) and (2) are easily solved for the expected monthly rates of flow of empty barges over each directed leg of the river:

$$M_{k,k+1} = \max\left[0, \sum_{j=1}^{k}\sum_{i=k+1}^{P} q_{ij} - \sum_{i=1}^{k}\sum_{j=k+1}^{P} q_{ij}\right]$$

$$M_{k+1,k} = \max\left[0, \sum_{i=1}^{k}\sum_{j=k+1}^{P} q_{ij} - \sum_{j=1}^{k}\sum_{i=k+1}^{P} q_{ij}\right] \text{ for } k = 1, \ldots, P - 1. \tag{3}$$

Minimum Frequency of Boat Service

Let B_{kh} denote the (deterministic) monthly rate of movement of boats over the directed leg from port k to adjacent port h; i.e., B_{kh} will be the number of boat trips made per month from port k to adjacent port h. Boats are assumed to follow a fixed schedule, following an itinerary through all ports in sequence. As was true in the case of barges, the availability of a finite number of boats and the assumed structure of the river imply, as an equilibrium condition, that the rate of flow of boats over a leg must be the same in each direction:

$$B_{k,k+1} = B_{k+1,k} \text{ for } k = 1, \ldots, P - 1. \tag{4}$$

A lower bound, denoted $\underline{B}_{k+1,k}$, for $B_{k+1,k}$ is given by

$$\underline{B}_{k+1,k} = \frac{1}{c}\left[\sum_{j=1}^{k}\sum_{i=k+1}^{P} q_{ij} + M_{k+1,k}\right] \text{ for } k = 1, \ldots, P - 1, \tag{5}$$

where c denotes the maximum number of barges permitted per tow. Of course $\underline{B}_{k+1,k} = \underline{B}_{k,k+1}$, for $k = 1, \ldots, P - 1$, as was the case with barges.

As will be shown later, the number of barges required to service a given load pattern is inversely related to the frequency of boat service. Consequently, there may be more than one *efficient*[2] combination of barge fleet size b and boat fleet B

---

[2] A barge and boat fleet (b, B) is said to be *efficient* if it is not possible to reduce either b or B while holding the other fixed and yet maintain statistical equilibrium of the system.

which will deliver the cargo of a given load pattern. Thus the possibility of boat service being more frequent than the lower bound just stated in equation (5) will be considered.

One might assume that the frequency of service over a leg is approximately proportional to the demand for service over that leg, subject to the lower bound constraint imposed by maximum flotilla size. It will simplify this discussion, however, to assume that the boats cross all legs with equal frequency. These two assumptions will be equivalent in case the demand for service is the same over all legs of the river district served. One instance in which such a case may be approximately realized is that in which a large proportion of the demand for transport is for cargo transit from one end point of the river district served to the other. Consequently the discussion here will be restricted to cases in which boats traverse the legs with equal frequency. It will be assumed that the boats turn around at the end points of the river district served (port 1 and port P) and also that boats wait a negligible amount of time in ports. As a result of this operating policy, no barges need spend time in intermediate ports.

Since a round trip for a boat requires

$$\sum_{k=1}^{P-1} (d_{k,k+1} + d_{k+1,k}) = RT \tag{6}$$

days, one boat can make $30/RT$ round trips per month. A fleet of B boats together make $30B/RT$ trips over every directed leg *per month*. It is assumed that the boats are spaced RT/B days apart so there will be RT/B days between the starts of consecutive boat trips over any directed leg.

However, a lower bound is imposed on the frequency of boat service by the load pattern and the limits on tow size. Since all legs are served equally frequently, the number of trips must be at least as great as the minimum requirements on the leg most heavily traveled by barges. That is,

$$30B/RT \geq \max \left[\underline{B}_{k+1,k} \text{ for } k = 1, \ldots, P-1 \right].[3]$$

Hence,

$$B \geq \frac{RT}{30} \max \left[\underline{B}_{k+1,k} \text{ for } k = 1, \ldots, P-1 \right]. \tag{7}$$

Approximate Barge Fleet Requirements

The number of barge days per month spent in transit is the sum over all directed legs of the product of the total number of barges to cross the directed leg multiplied by the time required to cross the leg. Since, recalling equations (1), the expected number of barges crossing a leg is the same in each direction, the expected monthly barge days in transit is equal to

$$\sum_{k=1}^{P-1} \left[\left(\sum_{j=k+1}^{P} \sum_{i=1}^{k} q_{ij} + M_{k,k+1} \right) (d_{k,k+1} + d_{k+1,k}) \right]. \tag{8}$$

[3] Recall that $\underline{B}_{k+1,k} = \underline{B}_{k,k+1}$.

Since, by hypothesis, there is no limit on loading and unloading facilities, the barges spend no time waiting to be loaded or unloaded.[4] The expected time spent in loading and unloading is

$$\sum_{j=1}^{P} \sum_{i=1}^{P} q_{ij} \, (d_i' + d_j'') \tag{9}$$

barge days per month.

A port of empty barge surplus (deficit) is a port at which more cargo terminates (originates) than originates (terminates). The empty barge surplus at a port k is denoted S_k and is defined by

$$S_k = \max \left[0, \sum_{i=1}^{P} q_{ik} - \sum_{j=1}^{P} q_{kj} \right] \text{ for } k = 1, \ldots, P . \tag{10}$$

It will be assumed that empties leaving a port of empty barge surplus are carried directly to the port of empty barge deficit at which they will be loaded. Under this assumption, the expected monthly number of empty barges to begin journeys is the sum over all ports of the empty barge surplus, namely $\sum_{k=1}^{P} S_k$.

It is not clear, however, that this policy is optimal with respect to minimizing expected delay to cargo awaiting loading. There may be a better operating rule under which some condition or set of conditions exists where an empty is retained for loading at an intermediate port and another empty sent on from the intermediate port to the destination (deficit) port at another time. In the latter case and under the assumption of equations (2), an upper bound for the expected monthly number of empty barges to begin journeys is $\sum_{k=1}^{P-1} (M_{k+1,k} + M_{k,k+1})$ and the lower bound is $\sum_{k=1}^{P} S_k$.

If barges become ready to begin their journeys according to the same probability distributions each day and if there is almost always room in a tow for any barge ready to enter it, then a typical barge ready to begin a journey may expect to wait $(1/2)\,(RT/B)$ days before departure from its port of origin, since boat departures over any leg occur every RT/B days. The expected waiting time of a barge ready to depart will be approximated by the quantity $RT/2B$.

If the travel times are not constant or if the boats are not equally time-spaced, then the waiting time of barges for movement may be greater. A similar waiting-time problem was considered by Takacs [1962]. He assumed that the probability of a bus arrival at a bus stop at any time t depended only on the

---

[4] This may be interpreted as indicated earlier in footnote 1.

elapsed time since the last bus arrival there. He then derived the average wait-ing time of a passenger arriving at the stop. Letting m denote the average inter-arrival time of the buses and V stand for the coefficient of variation[5] of the dis-tribution of bus interarrival times, Takacs' result for the average waiting time of a passenger may be written as $(m/2)(1 + V^2)$. Thus, if the interarrival time is constant, $V = 0$ and the average waiting time equals $m/2$; otherwise the aver-age waiting time is greater.

Under the assumptions made here, there are averages of $\sum_{k=1}^{P}\sum_{j=1}^{P} q_{kj}$ loads and $\sum_{k=1}^{P} S_k$ empties which begin journeys per month. Since boats turn around at end points only and spend no time in ports, barges need spend no time waiting in intermediate ports. Hence the approximate expected number of barge days per month spent in waiting for movement is

$$\left[\sum_{k=1}^{P}\sum_{j=1}^{P} q_{kj} + \sum_{k=1}^{P} S_k \right] \frac{RT}{2B}. \tag{11}$$

Thus far, approximations for the average numbers of barge days required per month in transit, loading, unloading, and awaiting movement have been given. In addition, provision must be made for barge days awaiting cargo if the system is to be viable. Even if there were no stochastic components in the barge-line's operations, there would still be a need to make this allowance, since it would be most unusual if the times of cargo arrivals and of barge availability could be made to coincide in every case.

Let b_k'' denote the mean number of barges awaiting cargo at port k, for $k = 1, \ldots, P$. The value b_k'' is functionally dependent upon the expected inter-arrival time $1 \bigg/ \sum_{j=1}^{P} q_{kj}$ of cargo (or of barges to be loaded) at port k. It may also be related to the interarrival time of boats, which bring loaded and empty barges, at port k; the less often barges arrive at a port the greater is the tendency of the barge availability times to cluster. Attempts to derive an explicit relation-ship between b_k'' and these other variables of the model were unsuccessful.

Cost Minimization

Bringing together the expressions for boat requirements and for barge re-quirements, the size and composition of the equipment fleet [i.e., the pair (B,b)] are to satisfy

---

[5] The coefficient of variation is the ratio of the standard deviation to the mean.

$$B \geq \frac{RT}{30} \max \left[\frac{1}{c} \left(\sum_{j=1}^{k} \sum_{i=k+1}^{P} q_{ij} + M_{k+1,k} \right) \text{for } k = 1, \ldots, P \right], \qquad (12)$$

and

$$b \geq \frac{1}{30} \sum_{k=1}^{P-1} \left[\left(\sum_{j=k+1}^{P} \sum_{i=1}^{k} q_{ij} + M_{k,k+1} \right) (d_{k,k+1} + d_{k+1,k}) \right]$$

$$+ \frac{1}{30} \sum_{j=1}^{P} \sum_{i=1}^{P} q_{ij} (d_i' + d_j'')$$

$$+ \frac{1}{30} \left(\sum_{k=1}^{P} \sum_{j=1}^{P} q_{kj} + \sum_{k=1}^{P} S_k \right) \frac{RT}{2B} + \sum_{k=1}^{P} b_k''. \qquad (13)$$

It is seen that the required barge fleet size depends in part upon the size of the boat fleet. This substitution relation between boats and barges depends upon the fact that the more frequently a boat passes a point, the less time will barges spend waiting.

Under the assumptions made here, the set of efficient fleets (B,b) for a given load pattern (q_{ij}) can be found by setting b equal to the right-hand side of relation (13); then the set is given by $[B, b(B)]$ subject to inequality (12). This set is efficient with respect to the various assumptions made; with the same data but different operating rules a different set might be obtained.

Let C_B denote the cost per month of a boat and let C_b stand for the cost per month of a barge. Then the minimum cost fleet (B^*, b^*) from among the alternatives presented is indicated by the tangency of the graph of the equation $b = b(B)$ mentioned above with the graph of the cost equation $C_B B + C_b b = C_T$, or by a boundary solution $[\underline{B}, b(\underline{B})]$ if this tangency occurs for a value of B less than $\underline{B}$. (See Figure 9.)

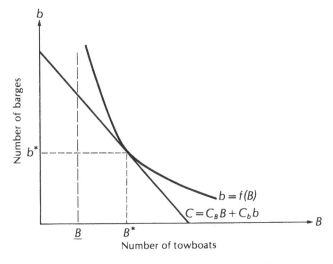

FIGURE 9.   Illustration of the optimum combination of boats and barges.

In a somewhat more general case, the expected delay time for cargo await-ing loading and/or movement may enter the cost minimization process. For each day that a shipment remains in the bargeline system in excess of the time required for loading, transit, and unloading, some cost may be incurred by the bargeline. Such cost may be partially or wholly implicit; to some extent it represents an inverse measure of the quality of the service of the bargeline and so would perhaps more properly be reflected in the revenue function. This cost could depend on the relation between the length of haul and the extra time this shipment was held. It will depend on the commodity, perhaps the shipper, and other systematic and nonsystematic factors.

Corresponding to an efficient fleet there will be some average time that barges spend waiting for cargo and some average time that cargo awaits barges for loading. The expected delay time for cargo awaiting loading can be reduced at the expense of greater barge waiting time by increasing the barge fleet beyond the minimum given by the right-hand side of relation (13). An explicit relation-ship between these variables has not been derived.

Let D denote the expected cargo delay awaiting loading. Let C_D denote the cost incurred for each day that a shipment waits for loading to begin, and let C_W stand for the cost for each day that a shipment waits, after loading, for move-ment. Since warehousing of some sort must be provided for cargo while it waits to be loaded, C_D may exceed C_W.

Total line-haul cost per month will be the sum of the cost incurred for the use of the equipment during the month plus the costs attributable to cargo de-lay and waiting. That is, expected cost per month will be

$$C_B B + C_b b + C_D D \sum_{j=1}^{P} \sum_{i=1}^{P} q_{ij} + C_W \frac{RT}{2B} \sum_{j=1}^{P} \sum_{i=1}^{P} q_{ij}. \tag{14}$$

Since b is functionally related to both B and D, only B need be chosen for any independently provided value of D.

ILLUSTRATIVE CALCULATION OF FLEET REQUIREMENTS
WITH VERIFICATION BY COMPUTER SIMULATION

A simple example will illustrate the calculation of approximate fleet re-quirements discussed above. Suppose a bargeline serves four ports which are linearly ordered. The expected monthly load pattern, in bargeloads, is as follows.

		Port of destination				Loads origi- nating	Net expected barge surplus	Net expected barge deficit
		1	2	3	4			
	1	0	4	1	46	51	10	0
Port of origin	2	10	0	3	2	15	8	0
	3	6	6	0	4	16	0	10
	4	45	13	2	0	60	0	8
Loads terminating		61	23	6	52	142	–	–

Row and column sums represent expected monthly bargeloads of cargo originating and terminating at each port. It is assumed that an average of three days is required for loading or unloading a barge at any port. That is, $d'_k = d''_k = 3$ for $k = 1, 2, 3, 4$. Transit times d_{kh}, in days, between adjacent ports follow.

		Port		
	1	*2*	*3*	*4*
Direction				
$\longrightarrow$		6	2	2
$\longleftarrow$		3	1	1

For example, travel time from port 2 to port 1 is three days. Travel time from port 3 to port 1 is four days. Summing, one finds that a round trip over the whole district served will require fifteen days. It is assumed that the maximum flotilla size c is sixteen barges.

From the expected load pattern, the expected number of loaded barges traversing each leg per month is found to be:

		Port			
	1	*2*	*3*	*4*	
Direction					
$\longrightarrow$		51	52	52	$\left(\displaystyle\sum_{j=k+1}^{P} \sum_{i=1}^{k} q_{ij} \right)$
$\longleftarrow$		61	70	60	$\left(\displaystyle\sum_{j=1}^{k} \sum_{i=k+1}^{P} q_{ij} \right).$

The difference between the loaded barge flow in each direction yields the empty barge flow necessary for equilibrium [see equations (3)]. The numbers of empty barges expected to cross each leg during a month are:

		Port		
	1	*2*	*3*	*4*
Direction				
$\longrightarrow$		10	18	8
$\longleftarrow$		0	0	0

These empty flows could also be determined easily from the barge surplus and deficit figures for each port. The expected total barge flow per month over each leg [see equation (1)] can be given without specification of direction of travel:

		Port		
	1	*2*	*3*	*4*
Barges		61	70	60

Dividing the rates of barge flow by the maximum number of barges (sixteen) which may travel together yields the minimum number of boat trips needed each month to carry the barges over each leg [see equations (5)]:

	Port			
	1	*2*	*3*	*4*
Minimum boat trips		$3^{13}\!/_{16}$	$4^{3}\!/_{8}$	$3^{3}\!/_{4}$

Under the assumption that boats make round trips only, the minimum boat requirements are governed by the requirements on the leg with the greatest barge traffic. Consequently, since a round trip requires fifteen days [using equation (7)], it is seen that the services of at least $(15/30)(70/16) = 2^{3}\!/_{16}$ boats will be required per month; i.e., $\underline{B} = 2^{3}\!/_{16}$ boats.

From equation (8), the expected number of barge days per month spent in transit totals $(61 \times 9) + (70 \times 3) + (60 \times 3) = 939$. And, from equation (9), the expected number of barge days per month spent in loading and unloading is $142 \times 6 = 852$. Thus, the expected number of barge days spent per month in transit, loading, and unloading is $939 + 852 = 1{,}791$.

According to (11), approximately $(142 + 18)(15/2B) = 1{,}200/B$ barge days per month will be spent in waiting for a towboat. The amount of barge time spent waiting for transit is inversely proportional to the frequency of boat service and hence to the number of towboats.

Let b' denote the right-hand side of (13) *less* $\sum_{k=1}^{P} b_k''$. Then $b' = (1/30)$

$(1{,}791 + 1{,}200/B)$. Since some number of barge days idle awaiting cargo for

loading is required for viability, $\sum_{k=1}^{P} b_k''$ is positive and therefore b' is necessarily

smaller than the number b of barges required altogether. Unfortunately, as was mentioned before, no good means has been found to estimate the number of barge days spent awaiting cargo. In the absence of this information, the relationship between B and b' is tabulated for some integral values of B.

B	b'
3	73.0
4	69.7
5	67.7

At least for this example, the extent of substitution possible between boats and barges appears slight relative to the comparative cost of the two types of equipment. But the increase in boat fleet from, say, three to four boats also results in the reduction of interdeparture time from five days to three and three-quarter days; this could be an important consideration with regard to cargo delay.

The relationship given between B and b' is only approximate. Further, since the analysis has been based only on expected values, we lack analytical results concerning the stochastic properties of the system's operations caused by the extent and form of variability in the realized load pattern and in loading and unloading times. It will also be recalled that the expected amount of time spent by cargo awaiting loading has not been included in the analysis, so that the effect of a marginal barge (in excess of the minimum number required for feasibility with a given boat fleet) on cargo waiting is not known. Finally, the values of b_k'', the mean numbers of barges awaiting cargo, have not been functionally related to the other characteristics of the system.

In an effort to develop an apparatus for experimental investigation of some of these questions, and as a beginning toward development of computer simulation as a tool in transportation equipment scheduling, a FORTRAN computer program was written to simulate bargeline line-haul operations. The program was run with data (parameters) of the example described and discussed above. The operation of the computer program for this example is sketched briefly below.[6]

A Simulation Program

The computer simulation model consists of a main program and two interchangeable cargo arrivals subroutines. The primary function of a cargo arrivals subroutine is to specify, according to an appropriate probability distribution, the amount of cargo arriving on each simulated day at each origin for each destination. Any cargo arrivals subroutine could be used with the main program. Those used here are described later.

Parameters read by the main program include those describing the numbers of barges initially located at each port. The mean time required for transit over each leg, the mean loading and unloading time required at each port, and the range of fluctuation about those means as a percent of the mean are also inputs for the simulator.

As each simulated day begins, the queues of cargo awaiting loading are incremented by the day's arrivals of new cargo at ports of origin. These arrivals are determined, as mentioned above, by a subroutine. Barges which are to finish loading or unloading are switched from the loading or unloading categories to the loaded-and-awaiting-movement and the empty classifications respectively.

The schedule of each boat is examined to determine which boats are to arrive in port on this day. For these boats, the next leg to be crossed is found from the given itinerary. (All boats were to traverse the river between ports 1 and 4.) The day that this next leg would be completed is found as the current date plus the mean transit time over the next leg to be crossed.

When a boat arrives at a port, any loaded barges that have reached their destination and *all empty barges* are removed from the tow. Barges in the former

---

[6] The FORTRAN program, with a guide to all symbols used, is available from the author, Nancy L. Schwartz, Graduate School of Industrial Administration, Carnegie-Mellon University.

category begin their unloading times. A completion date for unloading is calculated as follows. A random number is drawn from a rectangular distribution centered on the mean unloading time at the port in question and with a range equal to a specified per cent of that mean; this number is rounded to the nearest integer (except that at least one day is always required) and added to the current day to obtain the date of completion of unloading.

Loaded barges to continue with the boat stay in tow. As long as there is room in the tow, any loaded barge which is to begin its journey by traversing the leg that the boat is to cover next is put into tow. Then, if the tow is still not full and there are empty barges at the port (either generated there through previous unloading or delivered there), consideration is given to the inclusion of empty barges in the tow. Of the loads of cargo which have been generated to date, the number that must travel over the leg which the boat will cross next and the number of such loads that must travel over that arc in the opposite direction are found. The difference represents the cumulative number of empty barges desired over the leg. If this number is greater than the number of empties already assigned to cross the leg, then empties are put into tow until a limit is reached: either the tow becomes full, the stock of empty barges at the port becomes depleted, or the number of empty barges assigned to cross the leg reaches the computed desired number. Then the tow is complete and the boat is dispatched.

After all boats are under way for the day, cargo is loaded at each port as long as cargo and empties are available. At each port, cargo is assigned to barges in rotation among the possible cargo destinations. Loading completion dates are computed in the same manner as unloading completion dates.

At the end of the simulated day, the numbers of loads of cargo awaiting loading or of empty barges at each port are recorded. Then another simulated day begins.

Simulation Results and Suggestions for Further Research

In all computer runs, all of the barges are initially distributed among the ports as empties. The system fills out as cargo arrives at ports of origin, is loaded into these empties, etc., according to the procedure just outlined. The single exception is that during the first twenty days of operation, no empty barges are permitted to travel.

For purposes of comparison, runs were also made in which the interarrival time of cargo for any specific origin-destination pair did not vary and the realized cargo load pattern was exactly the same from month to month (see Table 10).

The cargo arrivals subroutine used with the data of Table 10 read the cargo arrivals during each of the first thirty days and then repeated those arrivals to the main program during each successive thirty-day period. In these runs, the travel, loading, and unloading times were taken as constant. Hence the existence

TABLE 10. Regular Cargo Arrivals at Ports of Origin through Month

Port of		Bargeloads
Origin	Desti- nation	
1	2	1 on days 8, 16, 22, 30
1	3	1 on day 15
1	4	3 on day 15; 2 on other odd days; 1 on even days
2	1	1 on days 3, 6, 9, 12, 15, 18, 21, 24, 27, 30
2	3	1 on days 10, 20, 30
2	4	1 on days 5, 20
3	1	1 on days 5, 10, 15, 20, 25, 30
3	2	1 on days 2, 7, 12, 17, 22, 27
3	4	1 on days 8, 16, 22, 30
4	1	1 on odd days; 2 on even days
4	2	1 on days 1, 3, 6, 9, 10, 12, 15, 18, 20, 21, 24, 27, 30
4	3	1 on days 5, 20

of an equilibrium was readily observable, as the daily lengths of cargo and empty barge queues were the same during each month after a few months.

The first sequence of computer runs was made with three boats and seventy-seven, seventy-eight, and seventy-nine barges, using the cargo arrivals subroutine described above and constant service times. The equilibrium behavior of the queues of cargo awaiting loading and of the empty barges at each port is summarized in Table 11, where the maximum and average daily queue lengths during a thirty-day period are shown. Table 11 also shows the numbers of days during a month that a given queue was not empty. With seventy-seven barges, the equilibrium pattern reported was first observed during month 10, while the runs with seventy-eight and seventy-nine barges reached equilibrium by the seventh month.

The data presented in Table 11 illustrate two important conclusions. *First, the equilibrium reached depends upon the initial distribution of barges. Second, the effects of an additional barge depend upon the port at which it is placed,* as well as upon the number and initial locations of the other barges. The effects are not evenly spread through the system.

Crude measures of the overall effect of an additional barge are the change in the sum over all ports of the mean cargo queue lengths and the change in the sum of the mean empty barge queue lengths. The addition of the seventy-eighth barge at port 4 resulted in a reduction of 1.8 in the mean number of loads of cargo awaiting loading and an increase of 1.2 in barges waiting in ports. Adding this barge at port 1 instead resulted in a reduction of 1.67 in the average number of cargo loads waiting and an increase of 1.0 in empty barges waiting. Comparing the second and fourth runs, it is seen that the seventy-ninth barge added at port 4 reduced the overall number of cargo loads waiting by 1.53 while the sum of barges waiting rose by 1.97. Adding the seventy-ninth barge at port 3 instead led to a cargo reduction of only 0.27 while the empty barges waiting rose by 1.07.

TABLE 11. Summary Thirty-Day Statistics for Simulation with Regular Cargo Arrivals and Three Boats

Run, no. of barges, and port	Initial barge dis- tribution	Cargo queue length		No. of days with non- empty cargo queue	Empty barges		No. of days with queue
		Maximum	Average		Maximum	Average	
		(...... loads ......)					
Run 1 (77 barges)							
Port 1	27	6	2.27	19	6	1.03	9
2	8	1	.53	16	4	.27	3
3	8	3	.87	18	1	.07	2
4	34	7	2.43	21	3	.43	7
Run 2 (78 barges)							
Port 1	28	5	1.80	18	7	1.23	10
2	8	1	.53	16	2	.27	6
3	8	2	.20	5	2	.73	16
4	34	6	1.90	20	4	.57	8
Run 3 (78 barges							
Port 1	27	6	2.13	19	6	1.07	11
2	8	1	.53	16	2	.27	5
3	8	2	.17	4	3	.87	17
4	35	6	1.47	16	4	.80	9
Run 4 (79 barges)							
Port 1	28	5	1.40	16	7	1.50	11
2	8	1	.53	16	2	.27	5
3	8	1	.03	1	3	1.73	26
4	35	5	.93	11	5	1.27	13
Run 5 (79 barges)							
Port 1	28	5	1.80	18	7	1.23	11
2	8	1	.53	16	2	.27	6
3	9	1	.03	1	3	1.73	26
4	34	6	1.80	20	4	.63	9

The second cargo arrivals subroutine used in the second sequence of computer runs generates the cargo arrivals internally on each simulated day. Loads are generated *in the system* at a mean rate of 142/30 loads per day; the actual number generated on a day is the integer closest to a number drawn from a rectangular distribution with end points equal to $(142/30) (1 \pm 0.3)$. The pair of ports representing the origin and destination is assigned to each load generated according to a fixed probability representing the relative frequency of the origin-destination in the total number of loads to be handled by the bargeline. The numbers of loads actually generated during each simulated month to originate at each port and to terminate at each port are recorded in the tables summarizing the simulations using this cargo generator.

Table 12 summarizes the operations of the simulation run with the load generator described above, using three boats, seventy-seven barges (initially distributed with twenty-seven, eight, eight, and thirty-four barges at ports 1 through 4 respectively), with all loading and unloading times chosen from a rectangular distribution ranging from one to five days, and with constant transit times. Table 13 is similar except that the four boats and seventy-three barges were initially distributed with twenty-six, seven, eight, and thirty-two barges at ports 1 through 4 respectively. Finally, Table 14 displays summary statistics for a run with five boats and seventy-one barges initially distributed with twenty-six, seven, seven, and thirty-one barges at ports 1 through 4 respectively. If a queue was positive on all thirty days of a month, then the minimum queue length during that month is shown in parentheses.

It appears that the sum of the b_k'' and the errors in the approximation of b' total between three and four with, as was expected, a higher amount applicable in the case of three boats and a lesser amount in the case of five boats.

On the basis of the simulation runs, it appears that the fleets used in the runs reported in Tables 12, 13, and 14 lie on or very near the efficiency frontier for the illustrative example. Speaking very loosely, an equilibrium may have been reached at some time between approximately the tenth and the twelfth months of simulated operation. The ranges of fluctuations of arrivals of cargo at ports of origin for ports of destination were rather large. Nevertheless, the very simple operating rules incorporated in the simulation program seem to have performed reasonably well, in terms of the amount of equipment required relative to the estimate of a lower bound, in terms of the range of fluctuation of queue lengths, and in terms of the infrequency of long queues of cargo awaiting loading. Further work is needed to produce operating rules that depend more heavily upon the current state of the bargeline system than do those in the present simulation program. Also, experimentation needs to be done for several cases with differing forms of the probability distributions over the cargo arrivals and the service times; in particular, it may be of interest to examine cases in which the probability distribution of unloading times, say, is skewed with a large upper tail.

As stated in connection with Table 11, the substitution relation between an incremental barge and the amounts of cargo awaiting empty barges needs to be explored further.

It has been assumed throughout this study that the transit time over any specified leg is constant. If transit times are stochastic, then the length of time barges wait for movement will tend to be greater than would be the case without this stochastic element, and consequently the need for transport equipment is greater. The extent of this effect constitutes another question for future exploration. In addition to the effect of random elements on transit times, it is known (see Chapter 2) that the speed of a tow varies systematically with the size of flotilla. A more complete study of equipment requirements and scheduling would incorporate this effect.

If the demand for barge transit varies considerably over the different legs of the river, an optimal boat policy might not involve all boats invariably making round trips. The saving in boat time resulting from a change in boat schedule must be weighed against any concomitant increase in barge time caused by

TABLE 12.　Summary Statistics for Simulation with Multinomial Cargo Arrivals During Thirty-Day Periods: Three Boats and Seventy-Seven Barges

Month	Total cargo awaiting loading			Number of empty barges			Cargo created during month to		Cargo delivered at port specified
	Maxi-mum	Average	Days with queue[a]	Maxi-mum	Average	Days with queue[a]	Origi-nate[b]	Termi-nate[b]	
	(..... loads .....)						(........ loads ....)		
Port 1									
1	0	0	0	27	13.37	(2)	57	55	38
2	2	.17	4	7	1.93	20	47	61	59
3	5	1.67	23	2	.20	4	57	54	57
4	7	3.90	(1)	0	0	0	55	49	50
5	8	2.47	23	4	.33	4	51	69	66
6	4	.57	7	10	3.00	23	53	59	63
7	5	1.10	16	5	1.20	12	58	63	61
8	5	.57	8	6	1.50	14	46	69	66
9	5	.67	8	11	3.10	19	50	61	58
10	5	.77	9	8	1.93	16	46	58	67
11	2	.07	1	10	2.70	24	43	61	58
12	2	.13	3	10	2.97	25	47	62	61
13	2	.07	1	9	4.50	28	41	62	60
14	3	.73	13	8	1.47	13	41	75	72
15	7	1.40	14	6	1.17	13	48	65	67
16	3	.60	10	4	1.20	15	46	61	61
17	8	1.93	14	7	1.40	14	57	60	61
18	7	1.37	16	6	1.17	10	45	61	68
19	12	3.67	21	5	.60	6	66	50	48
20	13	8.37	29	0	0	0	47	60	55
21	12	5.90	(2)	0	0	0	62	62	64
22	5	.53	8	6	1.73	17	NR	NR	61
Port 2									
1	2	.20	3	7	3.60	24	10	26	10
2	3	.57	9	4	1.00	12	16	30	37
3	0	0	0	9	2.63	27	12	26	23
4	1	.07	2	13	6.67	28	14	26	31
5	0	0	0	6	2.20	23	10	20	20
6	5	1.57	21	2	.07	1	21	21	20
7	6	2.63	25	3	.23	3	17	25	23
8	5	1.03	15	5	.53	9	14	24	29
9	3	1.10	17	3	.80	12	17	23	19
10	2	.13	3	3	1.00	17	17	28	27
11	2	.23	4	5	1.03	15	10	25	26
12	3	.80	16	2	.37	7	17	21	24
13	3	.50	9	2	.40	10	17	25	28
14	2	.73	17	1	.13	4	12	24	17
15	2	.37	8	3	.70	13	17	23	27
16	2	.53	11	2	.47	11	14	23	20
17	2	.13	2	7	2.93	19	7	19	21
18	5	2.17	20	3	.50	9	15	21	16
19	3	.20	3	7	3.43	26	14	20	26
20	0	0	0	10	7.40	(5)	15	18	18
21	3	.20	4	10	5.00	25	12	24	19
22	3	.33	4	4	1.53	18	NR	NR	29

Table 12—Continued

Month	Total cargo awaiting loading			Number of empty barges			Cargo created during month to		Cargo delivered at port specified
	Maximum	Average	Days with queue[a]	Maximum	Average	Days with queue[a]	Originate[b]	Terminate[b]	
	(..... loads .....)						(........ loads ....)		

Port 3

Month	Maximum	Average	Days with queue	Maximum	Average	Days with queue	Originate	Terminate	Cargo delivered
1	1	.07	2	7	4.03	27	16	11	7
2	7	2.53	20	4	.27	3	13	6	7
3	4	1.00	12	8	3.80	18	11	3	5
4	0	0	0	7	4.27	30	14	8	8
5	1	.20	6	6	2.50	21	6	7	4
6	5	1.13	10	5	1.73	18	10	7	10
7	1	.07	2	5	3.33	27	9	8	8
8	3	.77	12	5	1.07	10	14	5	4
9	6	.83	8	4	1.00	16	19	8	9
10	8	2.93	24	2	.20	3	16	10	9
11	3	1.10	16	5	1.33	12	15	8	8
12	6	2.63	20	3	.47	8	16	4	6
13	8	2.50	21	3	.23	4	17	5	3
14	10	1.77	16	1	.10	3	10	7	5
15	11	6.07	29	0	0	0	16	4	8
16	4	1.57	22	3	.23	3	15	11	8
17	5	1.43	12	9	3.40	18	14	6	8
18	3	.60	10	6	2.13	15	10	6	2
19	1	.17	5	7	1.37	17	12	6	8
20	5	3.20	(1)	0	0	0	18	11	12
21	6	2.93	29	1	.03	1	15	10	6
22	9	5.93	(2)	0	0	0	NR	NR	10

Port 4

Month	Maximum	Average	Days with queue	Maximum	Average	Days with queue	Originate	Terminate	Cargo delivered
1	12	3.80	16	30	7.57	14	66	57	28
2	11	4.67	23	4	.60	7	68	47	55
3	4	1.13	14	5	1.17	13	64	61	58
4	2	.13	3	6	2.27	25	54	54	56
5	7	1.33	14	8	1.73	14	78	49	47
6	6	1.47	12	4	.93	14	61	58	56
7	5	.73	11	6	1.63	13	64	52	51
8	6	1.10	13	8	1.27	13	67	43	51
9	1	.03	1	7	3.20	26	59	53	54
10	3	.70	9	7	2.03	18	64	47	43
11	6	1.00	10	9	2.57	18	70	44	51
12	0	0	0	12	6.20	28	57	50	39
13	0	0	0	11	5.50	(2)	61	44	52
14	7	2.63	19	5	.90	7	84	41	43
15	3	.57	11	6	1.37	14	63	52	39
16	7	1.30	11	7	1.77	17	67	47	58
17	6	1.53	14	10	2.37	14	68	61	53
18	10	3.87	23	4	.57	6	68	50	50
19	6	.57	7	10	3.70	22	53	69	62
20	1	.03	1	13	6.47	29	57	48	51
21	9	3.07	23	1	.13	4	64	57	60
22	7	2.73	19	2	.30	7	NR	NR	50

NR = Not recorded.

[a] If there was a queue every day, the minimum queue length during the month is shown in parentheses.

[b] Originating or terminating at specified port.

TABLE 13.   Summary Statistics for Simulation with Multinomial Cargo Arrivals During Thirty-Day Periods: Four Boats and Seventy-Three Barges

Month	Total cargo awaiting loading			Number of empty barges			Cargo created during month to		Cargo delivered at port specified
	Maximum	Average	Days with queue[a]	Maximum	Average	Days with queue[a]	Originate[b]	Terminate[b]	
	(..... loads .....)						(........ loads ....)		

Port 1

Month	Max	Avg	Days	Max	Avg	Days	Orig	Term	Deliv
1	1	.10	3	24	9.30	27	50	65	45
2	5	.83	11	5	.87	10	46	63	63
3	5	.87	11	4	.77	15	60	53	53
4	4	1.07	16	2	.47	9	54	64	61
5	5	1.23	17	4	.53	9	55	53	56
6	3	.80	15	6	1.50	13	44	68	62
7	4	1.03	12	7	1.43	17	51	56	66
8	6	2.03	22	2	.23	4	50	47	44
9	8	2.77	22	4	.40	6	50	65	61
10	7	1.87	21	3	.57	9	60	65	60
11	5	1.50	19	2	.37	8	55	62	66
12	8	2.97	24	5	.37	3	56	49	54
13	8	3.37	25	8	.50	3	53	64	61
14	10	3.03	20	3	.60	7	50	56	58
15	5	1.13	12	4	1.07	14	50	67	63
16	3	.50	9	7	1.37	15	40	79	70
17	5	1.23	15	4	.90	12	46	70	78
18	4	1.37	18	5	.53	8	45	56	60
19	3	.57	10	5	1.33	16	55	62	57
20	6	.97	13	7	1.07	14	49	64	67
21	5	1.37	13	8	1.53	13	39	68	70
22	5	.93	10	6	1.63	15	NR	NR	60

Port 2

Month	Max	Avg	Days	Max	Avg	Days	Orig	Term	Deliv
1	6	1.57	14	7	1.47	15	19	17	11
2	6	2.13	22	3	.23	3	14	21	20
3	1	.03	1	4	2.00	23	17	21	22
4	2	.40	8	7	2.13	17	8	25	23
5	0	0	0	7	2.73	22	11	24	27
6	4	.73	11	6	1.83	16	16	26	25
7	2	.43	10	4	.83	9	6	22	21
8	0	0	0	8	5.17	(3)	21	24	25
9	2	.13	3	11	3.63	24	14	23	23
10	4	1.57	24	2	.20	4	13	23	24
11	4	1.37	22	5	.73	7	11	20	20
12	1	.10	3	9	4.90	24	13	26	23
13	3	.50	10	7	2.97	19	17	21	24
14	1	.10	3	9	4.70	24	8	26	21
15	3	1.17	17	3	.33	6	18	20	24
16	12	6.57	29	0	0	0	24	18	19
17	14	9.60	(7)	0	0	0	19	20	15
18	8	3.83	25	1	.07	2	10	28	27
19	1	.07	2	5	1.53	18	13	25	31
20	6	2.40	22	4	.20	2	16	21	19
21	4	.87	15	1	.17	5	14	23	19
22	1	.23	7	4	1.30	15	NR	NR	31

TABLE 13—Continued

Month	Total cargo awaiting loading			Number of empty barges			Cargo created during month to		Cargo delivered at port specified
	Maximum	Average	Days with queue[a]	Maximum	Average	Days with queue[a]	Originate[b]	Terminate[b]	
	(..... loads .....)						(........ loads ....)		

Port 3

Month	Maximum	Average	Days with queue	Maximum	Average	Days with queue	Originate	Terminate	Delivered
1	3	.67	12	8	2.67	17	13	2	0
2	4	.77	9	6	1.37	17	9	7	9
3	1	.07	2	7	2.53	25	13	11	9
4	3	.40	7	4	1.97	20	9	2	4
5	1	.03	1	5	2.53	27	7	8	7
6	5	1.27	16	8	1.93	12	19	5	3
7	4	1.33	21	4	.60	7	18	8	10
8	7	2.00	17	4	1.43	13	15	7	4
9	7	2.13	18	3	.33	4	8	9	11
10	2	.17	4	4	1.77	22	11	6	7
11	3	.60	9	5	1.40	16	18	7	5
12	1	.30	9	7	2.20	17	8	6	8
13	0	0	0	8	3.77	19	6	8	4
14	3	.20	3	7	1.63	18	20	9	10
15	4	1.27	20	1	.17	5	18	7	10
16	4	1.50	17	1	.30	9	10	9	8
17	7	2.27	22	4	.53	8	19	8	7
18	8	4.77	(1)	0	0	0	16	12	14
19	5	1.77	19	1	.17	5	19	2	2
20	2	.37	6	6	2.10	20	12	4	3
21	6	2.47	21	2	.17	4	13	5	5
22	8	2.43	24	2	.23	5	NR	NR	7

Port 4

Month	Maximum	Average	Days with queue	Maximum	Average	Days with queue	Originate	Terminate	Delivered
1	5	1.07	9	30	7.53	19	59	57	36
2	1	.03	1	6	2.27	25	69	47	48
3	0	0	0	14	6.40	29	53	58	57
4	7	1.37	12	10	2.27	15	74	54	47
5	9	1.13	8	4	1.03	15	64	52	56
6	3	.27	4	6	2.13	22	65	45	53
7	4	.30	3	10	3.87	24	61	50	44
8	0	0	0	13	8.60	( )	47	55	52
9	10	2.40	15	11	1.87	11	75	50	51
10	7	1.30	12	6	1.10	12	66	56	56
11	1	.07	2	6	3.23	24	59	54	50
12	3	.23	3	8	2.17	24	53	49	59
13	3	.20	2	9	2.70	21	64	47	46
14	7	2.47	18	5	.93	12	64	51	51
15	3	.27	4	7	3.10	24	62	54	46
16	5	1.30	14	6	.77	10	76	44	47
17	9	2.10	18	4	.80	10	61	47	42
18	8	3.67	26	7	.40	3	68	43	45
19	0	0	0	11	6.17	(2)	56	54	57
20	7	1.47	13	7	2.27	16	65	53	55
21	7	1.37	13	8	2.00	15	68	38	41
22	6	.90	10	10	2.73	16	NR	NR	45

NR = Not recorded.
[a] If there was a queue every day, the minimum queue length during the month is shown in parentheses.
[b] Originating or terminating at specified port.

TABLE 14. Summary Statistics for Simulation with Multinomial Cargo Arrivals During Thirty-Day Periods: Five Boats and Seventy-One Barges

Month	Total cargo awaiting loading			Number of empty barges			Cargo created during month to		Cargo delivered at port specified
	Maxi-mum	Average	Days with queue[a]	Maxi-mum	Average	Days with queue[a]	Origi-nate[b]	Termi-nate[b]	
	(..... loads .....)						(........ loads ....)		
				Port 1					
1	1	.10	3	24	6.37	24	53	57	41
2	2	.27	5	5	1.67	22	47	56	56
3	6	1.23	11	6	1.77	17	57	52	53
4	7	3.27	25	2	.13	3	59	68	65
5	4	.93	11	8	1.90	14	51	56	58
6	4	1.03	13	4	1.07	12	52	55	54
7	4	1.00	13	8	1.37	15	37	74	66
8	5	1.13	17	5	.40	6	53	69	74
9	7	2.30	24	3	.20	3	57	64	69
10	5	.63	8	4	.93	15	51	51	51
11	3	.37	5	4	1.13	19	48	50	49
12	2	.27	6	5	1.10	14	52	58	60
13	7	1.53	13	7	1.97	17	45	69	58
14	6	1.13	12	5	.87	11	51	56	65
15	3	.77	13	6	.93	13	51	54	54
16	9	1.63	11	7	1.83	14	50	60	64
17	3	.50	10	7	1.70	14	50	69	63
18	3	.37	8	5	1.27	13	42	64	62
19	3	.70	14	4	.63	10	51	66	66
20	6	1.70	19	4	.83	9	45	68	67
21	3	.37	8	6	1.50	17	42	67	64
				Port 2					
1	1	.07	2	11	5.67	27	10	34	19
2	1	.07	2	4	1.10	15	13	22	32
3	2	.17	4	7	3.73	23	9	21	18
4	2	.33	8	8	2.53	21	13	15	17
5	0	0	0	6	2.07	23	19	23	24
6	1	.03	1	7	3.57	21	9	21	20
7	7	2.20	24	2	.20	4	17	23	25
8	12	6.53	(2)	0	0	0	19	19	15
9	4	1.23	15	6	.80	7	14	17	22
10	2	.23	5	5	1.40	15	13	34	24
11	2	.13	3	8	4.13	26	13	29	37
12	4	1.23	19	6	1.03	10	15	25	27
13	2	.60	11	4	1.17	15	12	25	19
14	7	3.50	27	2	.13	3	11	20	27
15	1	.03	1	7	3.43	27	7	33	26
16	4	1.23	20	4	.37	5	20	31	33
17	2	.43	10	3	.53	11	10	21	24
18	3	.37	8	3	.60	11	14	27	24
19	3	1.33	23	4	.37	5	16	25	22
20	6	2.20	20	4	.53	9	21	20	21
21	6	3.20	27	2	.13	2	15	17	21

TABLE 14—Continued

Month	Total cargo awaiting loading			Number of empty barges			Cargo created during month to		Cargo delivered at port specified
	Maximum	Average	Days with queue[a]	Maximum	Average	Days with queue[a]	Originate[b]	Terminate[b]	
	(. . . . . loads . . . . .)						(. . . . . loads . . . .)		

					Port 3				
1	10	2.50	14	7	1.37	13	19	8	4
2	10	3.27	24	1	.03	1	20	10	11
3	1	.03	1	5	2.33	25	11	3	5
4	2	.13	3	5	2.63	25	11	6	7
5	3	.53	10	5	1.30	18	14	6	6
6	1	.07	2	5	3.03	24	11	7	4
7	6	1.63	22	1	.03	1	16	6	6
8	8	4.03	21	4	.50	6	13	4	5
9	3	.80	16	3	.73	13	15	7	4
10	3	.83	13	4	.97	15	16	8	13
11	4	.70	10	5	1.00	14	18	8	6
12	0	0	0	5	1.83	20	10	5	4
13	4	1.53	19	2	.47	10	11	6	9
14	7	3.50	27	2	.13	3	18	9	6
15	1	.03	1	4	1.50	19	8	8	8
16	5	1.47	13	3	.37	7	17	7	7
17	1	.10	3	5	1.63	18	6	4	6
18	5	1.63	18	6	1.27	9	9	8	8
19	6	2.33	18	3	.47	8	16	3	3
20	4	.87	13	3	.87	14	16	6	3
21	5	1.67	21	2	.20	5	12	10	14

					Port 4				
1	9	2.07	14	27	7.10	15	66	49	33
2	5	.93	12	8	2.03	15	61	53	51
3	2	.17	4	6	2.33	23	59	60	57
4	5	1.07	14	4	1.00	13	61	55	49
5	4	.17	2	5	2.00	23	60	59	62
6	7	1.00	11	4	1.40	15	68	57	61
7	0	0	0	9	3.60	28	71	38	38
8	10	1.73	12	6	1.60	14	62	55	46
9	12	3.83	23	7	.73	6	59	57	60
10	8	1.73	14	7	1.50	15	66	53	59
11	1	.03	1	8	4.10	28	61	53	53
12	6	.60	6	7	1.90	18	67	56	54
13	4	.17	2	9	3.20	24	77	45	44
14	13	3.20	16	2	.20	5	63	58	52
15	4	.80	14	7	1.57	13	74	45	56
16	2	.07	1	8	3.33	26	56	45	39
17	1	.03	1	8	3.37	26	72	44	48
18	5	1.30	17	4	.87	9	77	43	44
19	8	1.47	12	4	1.20	15	63	52	52
20	3	.43	6	9	3.47	22	64	52	54
21	0	0	0	6	3.13	26	66	41	38

[a] If there was a queue every day, the minimum queue length during the month is shown in parentheses.
[b] Originating or terminating at specified port.

barges waiting in intermediate ports for boats. If the bargeline operates over a river with branches, or with differing maximum tow sizes over the various reaches of the river, further attention must be given to the relationships between alternative boat schedules and the waiting time of barges for movement both in ports of origin and in intermediate ports.

Finally, we note that in the bargeline industry interrelationships between carriers may materially affect equipment requirements. It has been assumed throughout this study that a firm is self-contained; i.e., that it is limited to using its own stock of equipment only. In fact, however, carriers lease barges to and from each other under short-term as well as long-term agreements. Further, other carriers may tow for them and they may tow for others. One effect of the possibility of making these short-term arrangements may be to smooth out the operations of a bargeline. Thus, if there should be an extraordinarily large amount of cargo to be moved from some point, then additional equipment might be leased from others to ease the cargo waiting. Similarly, a bargeline may be able to use productively, via charter to others, equipment that would otherwise be idle, moved empty, or underutilized (in the case of boats) because of unusually slack demand.

CONCLUSIONS

The complexity of a transport system and its stochastic nature combine to make analytic determination of optimum fleets and schedules very difficult. It appears, however, that approximate analytical results of the type obtained in this chapter can provide general guidelines for equipment requirements. These can then be checked out in greater detail through computer simulation of the operations of the bargeline. Computer simulation permits experimentation not only with sizes of fleets of barges and boats, but also with scheduling rules and boat itineraries.

5

OPTIMUM TRAFFIC FLOW, CONGESTION, AND DESIGN IN A WATERWAY SYSTEM: DETERMINATION BY SIMULATION

In this chapter, computer simulation is applied to the problems of optimal design and control of inland waterway transportation. The model describes the system, including operating policies, in a way that appears to be realistic enough to be of use to the Corps of Engineers in making benefit-cost evaluations of alternative waterway improvements. Its main advantage is that it permits detailed analysis both of congestion costs at all points of the system and of their interrelationships. The objective is to minimize the total cost of transporting goods through a waterway system.

Systems analysis[1] has been used in the past in analyzing inland waterway systems, particularly the Welland Canal of the St. Lawrence Seaway,[2] but computer simulation has only been used in analyzing the performance of river systems to the extent of analyzing the performance of a single lock [Carroll, 1968]. The present model demonstrates the inadequacy of an analysis of a single system component; e.g., a lock. It shows that when the system is congested, structural improvements or changes in operating rules at a single component may reduce delay costs there but transfer congestion to other parts of the system. It is the reduction in delay cost for the entire system which is relevant for benefit-cost analysis.

Another form of systems analysis which incorporates interrelated elements has been used in the past by the Board for Rivers and Harbors of the Corps of Engineers in analyzing inland waterway systems. It consists of simulations by hand without stochastic elements. An ingenious method of graphic analysis has been developed by personnel at the Board [Shultz, 1966]. The advantages of a computer model over the graphic approach lie in the systematic incorporation of the stochastic elements of the operation of a waterway system (which will be shown to be important) and the speed and low cost of production and operation of a computer model. This, in turn, leads to great flexibility in experimenting with alternative systems, traffic patterns, and operating rules.

A third form of systems analysis which has been applied to waterway systems is the analytic queuing model [e.g., Lave and DeSalvo, 1968]. The ana-

---

[1] Systems analysis can be defined as "the explicit quantitative analysis which is designed to maximize or at least increase the value of objectives achieved minus the value of resources used" [Hitch, 1967]. The value of resources used consists here of the private transport costs borne by commercial barge operators and the public costs of waterway infrastructure, especially the costs of improvements in locks or channels.

[2] A canal in which all traffic goes completely through the waterway differs substantially from a river with many intermediate ports.

73

lytic models developed to date may not appear sufficiently realistic to describe waterway systems adequately, but such models can produce accurate predictions of waiting time at individual locks. If the locks of a system are far enough apart to be treated as independent, then systems predictions from the analytical models may be very good. The operation of the system under conditions of congestion and nonindependence of locks can be described only by conditional probabilities dependent on the state of the system, and available analytic models either are not state-dependent in nature or incorporate assumptions of statistical independence among components of the system that have been clearly contradicted by observations on certain systems [Harris, 1967].

In congested systems, odd things can happen. Queues can transfer from one point (e.g., a lock) to upstream or downstream points as a result of random events or small changes in operating parameters. Local optimization of rules for the passing of individual tows through the locks may be inefficient in terms of the operation of the total system. Most analytic queueing models also lack flexibility in terms of the types of service time and arrival distributions that can be handled. Such distributional forms have marked effects on system operation and resultant congestion costs, as has been demonstrated with the present computer model. Data presented herein clearly show that the assumption of an exponential distribution does not agree with data collected by the Corps of Engineers on locking operations on the Ohio River (see Table 16, page 80).

The advantages and disadvantages of simulation models *versus* analytical models have been very crisply described by Lester B. Lave:[3]

> Simulations and analytical models of locking have very different properties. Analytical models are necessarily simpler and easier to understand. One can gain a deep understanding of the structure of the model and how changes in the assumptions affect conclusions. These models are simple enough so that one can have confidence that one understands them. The principal difficulty is that the assumptions may have little correspondence with reality. This fact itself is not sufficient to disqualify them: after all, we are not interested in incorporating reality in the assumptions, but rather in having the conclusions correspond to reality. Thus, the test of an analytical model is how well it predicts queues in practice—by the way, the Lave-DeSalvo model did exceedingly well on the Illinois waterway. The advantage of simulations is that they can explicitly incorporate more diverse and difficult problems. One can build in any set of assumptions that seems reasonable—but note that one will never correspond completely to reality, only the real system does that. The difficulty with simulations is that they become unwieldy. One has no good idea what causes what and to which parameters the model is sensitive. In some sense, one is trading plausibility of the assumptions for the ability to grasp the model as a unit.

THE COMPUTER MODEL

The two main topics that will be discussed here are: (1) the determination of an optimum volume of traffic on existing waterways, and (2) the determination of a socially optimum program of public investment in each waterway

[3] Correspondence with Howe dated October 21, 1968.

segment and the form that this investment should take. As with all activities which combine privately provided inputs with a publicly provided capital facility (e.g., road and air transport, public recreation facilities, etc.), there is a problem of co-ordinating public and private decisions to obtain economically efficient use of the facility. Under present policies, private users generally do not pay for the use of the capital facility.

A second aspect of the waterway problem that resembles other problems of public investment is the existence of large stochastic components in the performance of the system. Furthermore, the major variable of interest in many problem areas—delays caused by congestion—may appear to be removed in space and time from the particular user units whose operations cause the delay. Thus it usually is not possible to attribute deterministic delay time and cost to particular units. These costs can be defined only as averages over time. Through time averaging within the computer program, it is possible to assign time averaged marginal delay costs to increments in certain classes of traffic.

The goal of the analysis is efficiency through equating marginal benefit with cost. Equity issues of the distribution of benefits and costs to users of the waterway, taxpayers, etc., are not addressed. Nor are other sectors of the economy brought into the analysis. The demand for inland waterway services is given exogenously to the model. It may well be that truly optimum solutions to waterway transportation problems lie in such actions as co-ordination with other modes of transport or control of plant location decisions and land use patterns. The present model does not incorporate such alternatives.

The computer model permits us to analyze the above problems for a portion of an inland waterway system large enough, presumably, to incorporate the externalities of congestion. The characteristics of the traffic (arrival rates into the system, flotilla configuration, horsepower, draft, itineraries, etc.) can be described by frequency distributions derived empirically and introduced into the model. The characteristics of the locks, lock approaches, and lock size and operating times can similarly be introduced. Characteristics of the channels (average width, depth, and current velocity) and characteristics of constricted passing areas can be described from empirical knowledge. Additional inputs—such as the operating rules at the locks, a cost function for tow operating costs, and a speed function for tow speeds—permit realistic problems to be analyzed by computer simulation.

The simulation model characterizes an inland waterway transport system containing up to ten locks (each with one or two chambers) and twenty ports. Some features of the model are particularly adapted to contemporary shallow water pusher-type technology. Tows which enter the system are processed through the locks, channels, and narrow delay points in the channels in accordance with a distribution of itineraries. Figure 10 is a schematic drawing of a simple three-lock, four-port system which was used in many of the test runs. Most of the test runs reported herein are modeled after a particular stretch of the Ohio River: the locks are Meldahl, Markland, and McAlpine and the ports, Cincinnati, Lawrenceburg, Madison, and Louisville (see Figure 11). However,

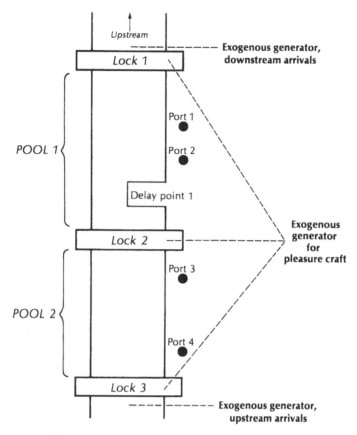

FIGURE 10.   Illustrative segment of a river system.

many of the parameters used in the test runs are estimates of actual distributions which can be improved when data collected by the Corps of Engineers at these locks are analyzed and available.

Arrivals of tows at the end points of the system (Locks 1 and 3 in Figure 10) are generated by a pseudo-random process based on the Poisson distribution, which is assumed independent of events within the system. The order of passage through the locks and ports is controlled by an itinerary assigned to each tow as it enters the system. An example of this might be entry at Lock 1, passage downstream through Lock 2 to Port 4, back upstream through Lock 2, and exit upstream at Lock 1. This particular trip traverses a constricted reach of channel labeled Delay Point 1 in our example. In this area, attainable tow speed may be different from that experienced in the rest of the channel because of channel restrictions. It may be assumed to be a no-passing area so that delays may be caused by a prior tow traveling in the opposite direction or a slower tow moving in the same direction.

In the model, tows interact at the locks in various ways. One type of interaction occurs when a tow bound in one direction approaches a lock that is processing another tow bound in the opposite direction. The second tow to arrive must wait at some designated distance below (or above) the lock until the first

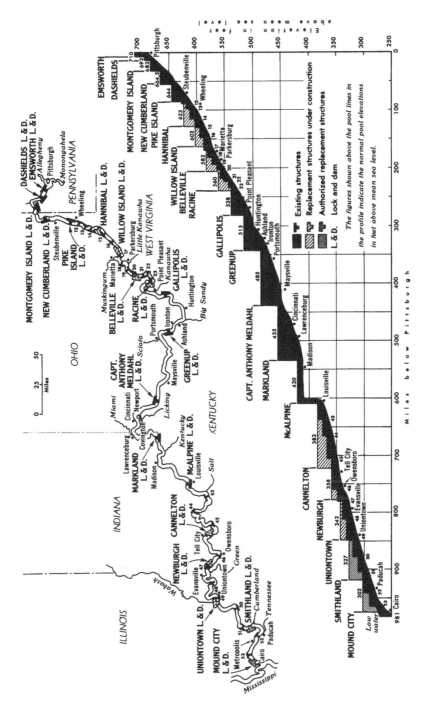

FIGURE 11. Navigation map of the Ohio River. (*Map based on U.S. Army, Corps of Engineers, Ohio River Division, General Plan, January 1966*)

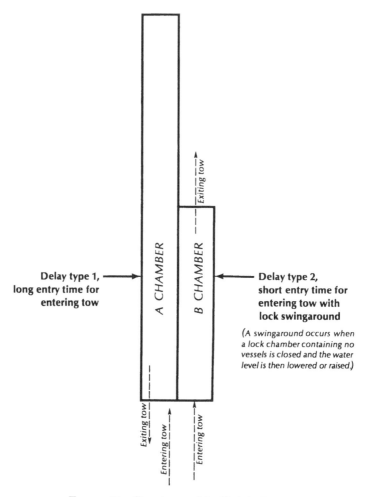

FIGURE 12.   Two types of traffic interference.

tow has completed its exit. If it had been following a tow traveling in the same
direction, the time required for entry would have been shortened because it
could then have proceeded to the lock wall to wait for the first tow to complete
its exit. These situations are illustrated in Figure 12. Thus, "short entry time"
or "long entry time," being a component of the total processing time at the
lock, is also treated as part of the delay time caused by congestion.

Point-to-point travel time between the ports and locks is based on distance
(established by inputs to the program), and on tow characteristics. In the example of Figure 10, the distance from Lock 1 to Port 1 is set at 35 miles; Port 1
to Port 2, at 20 miles; and Port 2 to Lock 2, at 45 miles. The average current
speed in this pool of 100 miles is 1.91 miles per hour. The average depth and
width are respectively 14.5 and 550 feet. The average speed at which a tow
travels in this pool is determined by the Howe equilibrium speed model [Chap-
ter 2] and is a function of the above channel characteristics and the width,
length, draft, and horsepower of the particular tow. The characteristics of the

TABLE 15.   Typical speeds of tows

Horse-power	Width	Length	Draft	Speed of tow in Pool 1		Speed of tow in Pool 2	
				Down-bound	Upbound	Down-bound	Upbound
	(.......... feet ....)			(................. mph ....)			
2,000	105	585	8.5	7.8	3.7	8.8	3.7
3,200	105	1,170	8.5	7.9	3.6	9.0	3.6
2,000	105	585	1.5	10.9	7.0	11.8	6.8
3,200	105	1,170	1.5	11.0	7.0	11.9	6.9

tow as it enters the system are determined by a random draw based on distributions of observed tow characteristics. Table 15 shows typical speeds for tows with varying characteristics, traveling upstream and downstream, in Pool 1 and Pool 2. The latter has been assumed to have a faster current speed and greater average channel depth and width.[4] A draft of 8.5 feet is associated with fully loaded barges, 1.5 feet for empty barges.

Figure 12 shows in schematic form how the operations of a lock are conceptualized in the model. Some locks currently in use have two chambers, the 110- by 1,200-foot A chamber, large enough to accommodate most tows in one pass, and the 110- by 600-foot B chamber, handling the smaller tows and pleasure craft, etc. A large tow can be processed through the B chamber by breaking the barge flotilla into two or more sections, each section being locked in turn. The computer model can utilize five distributions of locking time, each relating to different ways in which tows of various sizes may be processed through the lock.

In the absence of queues at the locks, the time taken to process the tow through the lock is conceptualized as consisting of the three random components shown in Figure 13: entry time, locking time, and exit time. During short entry time, the tow maneuvers from the lock wall into the lock chamber. Locking time (depending on the relation of tow size to chamber size) is the time taken to close the lock gates and fill or empty the loaded chamber so that the tow is in position to exit from the chamber. During exit time, the tow maneuvers out of the chamber and clears the lock area.

Swingaround (i.e., filling or emptying the lock chamber to accommodate a tow from the opposite direction) is, in the absence of queues, presumed to take place before the tow arrives at the lock wall. In this case, swingaround time has no effect on the time taken to process tows through the lock except when the operation is included in double or triple lockings.

Delays caused by congested operations at the lock are of two types. In Type 1 delays, a tow is delayed, as already described, by a prior tow coming

[4] Generally, river currents are more harmful than beneficial to tow speed. Partly this is reflected in the "slope-drag" factor, but there is an adverse effect on maneuverability which results in a greater loss of speed upstream than is gained in going downstream.

TABLE 16. Empirical Frequency Distributions of Components of Locking Time (Minutes), Meldahl Locks, Ohio River

Short entry time values

	1	2	3	4	5	6	7	8	9	10	11	12+	Means
Downbound relative frequency	.214	.331	.192	.078	.085	.028	.032	.018	.004	.007	.004	.007	3.0
Upbound relative frequency	.258	.277	.235	.072	.076	.023	.015	.011	0	.011	0	.023	3.0

Long entry time intervals

	[0,7]	(7,10]	(10,12]	(12,14]	(14,16]	(16,18]	(18,20]	(20,22]	(22,24]	(24,26]	(26,28]	(28+]	Means
Downbound relative frequency	.038	.192	.154	.308	.038	.115	.077	.038	.038	0	0	0	13.7
Upbound relative frequency	.030	.273	.182	.030	.121	.091	.030	.030	.061	.030	.030	.091	16.0

Exit time intervals

	[0,2]	(2,4]	(4,6]	(6,8]	(8,10]	(10,12]	(12,14]	(14,16]	(16,18]	(18,20]	(20+]	Means
Downbound relative frequency	.007	.098	.339	.205	.228	.068	.020	.020	.013	.003	0	7.6
Upbound relative frequency	.007	.084	.330	.242	.253	.051	.013	.010	0	.010	0	7.5

Swingaround time values

	[0,9]	10	11	12	13	14	15	16	17	18	19	20+	Means
Downbound relative frequency	.032	0	.065	.645	.194	.032	.032	0	0	0	0	0	12.2
Upbound relative frequency	0	0	.033	.467	.367	.067	.067	0	0	0	0	0	12.7

Locking time, A chamber

	(0,14]	(14,17]	(17,20]	(20,23]	(23,26]	(26,29]	(29,32]	(32,35]	(35,38]	(38,41]	(41,44]	(44+]	Means
Downbound relative frequency	.004	.002	.182	.212	.249	.138	.097	.041	.026	.015	.011	.004	25.2
Upbound relative frequency	.007	.030	.075	.187	.225	.165	.127	.090	.049	.015	.007	.022	27.1

Locking time, B chamber

	[0,14]	(14,16]	(16,18]	(18,20]	(20,22]	(22,24]	(24,26]	(26,28]	(28,30]	(30+]	Means
Downbound relative frequency	.057	.143	.257	.171	.057	.086	.086	.057	0	.086	21.1
Upbound relative frequency	.087	.130	.217	.087	.217	.130	0	.043	0	.087	20.5

Locking time, exponential[a]

	(0,14]	(14,17]	(17,20]	(20,23]	(23,26]	(26,29]	(29,32]	(32,35]	(35,38]	(38,41]	(41,44]	(44+]	Means
Downbound relative frequency	.429	.059	.058	.051	.046	.040	.036	.032	.028	.025	.022	.174	25.2
Upbound relative frequency	.405	.062	.056	.050	.044	.040	.036	.032	.028	.026	.023	.198	27.1

[a] Exponential relative frequencies with the same means and intervals as those of the A chamber empirical frequency distributions.
Source: Study of locking operations carried out at Meldahl Locks in the summer of 1966.

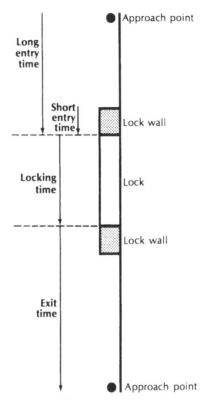

FIGURE 13.   Components of locking time.

through the lock in the opposite direction. In this type of delay, the difference between long entry time and short entry time is recorded as part of the delay at the lock. Type 2 delays occur when a tow must wait for a prior tow proceeding in the same direction to finish passing through the chamber. Under these conditions, the swingaround time for the empty chamber is part of the delay experienced by the second tow. The swingaround is assumed to begin when the chamber is empty and does not await the completion of the entire exit operation of the prior tow.

The time taken by each component of the locking operation is treated as a random variable. The distributions will, in most cases, be different for tows traveling upstream and downstream. Therefore, each component of locking time for each lock, upstream and downstream, is provided to the simulation program as a probability distribution with differing ranges and relative frequencies. It is assumed that these distributions are independent of tow configuration.[5] Table 16 shows empirical frequency distributions made up for the Meldahl lock, based on records kept by the Corps of Engineers over a two-month period in the summer of 1966. Also shown are the distribution values which would be implied by an

[5] Alan Chandler of the Corps of Engineers' Ohio River Division has stated that entry and, possibly, locking times may be functions of tow size and other characteristics. Data currently available should permit a test of this hypothesis. If it is borne out, minor program modifications can incorporate it in the model.

exponential distribution with the same mean. Formal queueing models usually assume exponential service times. Differences between the two are obvious and might have an important influence on the total amount of delay experienced in systems operations. This tends to cast some doubt on the relevance of analytic models which assume exponential service times.

The rule by which the chamber to be used is selected is part of the program and can easily be changed. As the program now stands, the first tow to arrive has locking priority and the chamber is selected on the basis of minimum *expected* time for the total locking operation (i.e., by using the mean times for the components of the entire locking operation as the lock master might do). Although the chamber selected is the one with the earliest expected completion time, the expected values used in this selection process are read in separately from the corresponding distributions. "Penalties" or "rewards" may be assigned to certain operations to influence the chamber selection. In this way, effectively different operating rules can be tested without rewriting the computer program.

The simulation program traces each tow from the time it arrives in the system to the time it finally departs. A clock is advanced from the recording of one event to the next subsequent event in time. The time that elapses between the entry and exit of each tow depends on a number of random endogenous events, namely the random times of locking operations and possible delays resulting from conflict in the use of the bottleneck facilities—the locks and constricted channel areas. In the operation of the program, after an appropriate warm-up period to allow the system to approach steady-state conditions, statistics are recorded on operations at the locks and at the delay points, both for the system as a whole and for each tow as it passes through the system. The program must be operated over a period of time long enough to permit the average value measures of performance to be reliable within desired limits of accuracy.

The major elements of the waterway transport simulation are summarized in Appendix A to this chapter. The elements of the simulation are characterized as: *components*, the conceptual elements of the system; *attributes*, the properties associated with components; *functions*, which essentially express mathematical or logical relationships among the attributes; and *events* in the action portion of the simulation where history is recorded either by changes in the numerical values of the attributes or in the creation or destruction of a temporary component.

Some further insights into the working of the simulation may be provided by Figures 14 and 15. Figure 14 is an optional computer program output tracing the unfolding of simulated events over time. For example, the events marked (1) in this figure trace the travels of Tow 2 downstream from Lock 1 to Port 4 to Lock 2. The event marked (2) shows the first instance of a delay caused by congestion at Lock 1.

Figure 15 is an example of the final computer program output. Most of the information shown there is self-explanatory. The total time the system was operated after the warm-up period was typically 44,000 minutes, approximately 30.6 days of calendar time. The columns headed "Time lost by tows passing through" refer to the delays experienced by tows about to use the A or B chamber and caused by prior operations in the chamber. A special report is made on multiple lockings (e.g., 14 double lockings for tows passing downstream at Lock 1). This gives additional information on a type of delay due to congestion, because in the absence of congestion these tows would have been single locked through the A chamber. However, the extra time spent in double locking is not compiled as delay time. The appropriate information for the evaluation of benefits and costs is shown below the figure. Total system cost of 1.45 million dollars represents private operating costs for a system performance of 1.98×10^9 gross ton-miles of traffic.[6] It might be noted that the ratio of gross ton-miles to net cargo ton-miles is probably in the range of two to three. Thus the cost per gross ton-mile of 0.7 mills would translate into a cost of 1.4 to 2.1 mills per net cargo ton-mile.

CRITERIA FOR JUDGING SYSTEM PERFORMANCE

In this section an attempt is made to establish criteria for judging system performance; that is, for concluding whether the system is performing better or worse after some structural or operating rule change has been introduced.

When marginal system cost equals marginal benefits, the waterway is being used optimally in terms of the average number of arrivals per day. When this level has been determined, it can serve as a benchmark for system performance. These cost and benefit terms require clear definition. It is assumed that variable public operating costs for the system are zero; that is, once the decision is made to keep the system open for navigation, all public operating costs are fixed and independent of traffic level. Opportunity costs of other system outputs (e.g., irrigation water) are ignored. The costs which vary with traffic level are private operating costs and can be partitioned into two classes:

- Straight-through operating costs; i.e., the costs that would be incurred by tows if they could proceed *without delay* in carrying out their itineraries. This would mean no channel congestion caused by other tows and locks ready to accept the tow every time locking is required.
- Delay-time costs which occur when a tow is kept waiting because of interference with other units, either in the channel or at the locks. (In the simulation runs reported herein, delays occur only at locks.)

Benefits should be measured through the use of the demand function which relates the average arrival rate into the system to the level of costs experienced by units using the system. The demand function should represent the willingness of users to pay for the use of the waterway; it should therefore represent, for every traffic level, the marginal willingness to pay *net* of straight-through op-

---

[6] It should be noted that tonnages used in the model are gross tons of 2,000 pounds, i.e., the weights of the barges are included in the measure. The justification for this is that the transfer of empty equipment among points of traffic imbalance also represents useful output of the system.

Tow number (downstream 1-99; upstream 200+)	No. of barges	HP of boat	Draft (feet)	Configuration (no. of barges across)	Clock time (minutes from program start)	Event	Processing time (minutes)	Idle time (minutes)	Time of completion (minutes from program start)
(1) 1	8	2,000	8.5	3	228	Arrival at Lock 1	43	0	271 B chamber
(1) 1	8	2,000		3	271	Exit from Lock 1	770	0	1,041 At Lock 2
2	17	3,200	8.5	3	542	Arrival at Lock 1	43	0	585 A chamber
2	17	3,200		3	585	Exit from Lock 1	266	0	851 At Port 1
3	8	2,000		3	693	Arrival from Lock 1	51	0	744 B chamber
(2) 4	17	3,200		3	698	Arrival at Lock 1	38	0	736 A chamber
5	17	3,200		3	705	Arrival at Lock 1	44	36	785 A chamber
4	17	3,200	8.5	3	736	Exit from Lock 1	418	0	1,154 At Port 2
3	8	2,000	8.5	3	744	Exit from Lock 1	269	0	1,013 At Port 1
201	8	2,000		3	751	Arrival at Lock 3	52	0	803 A chamber
6	8	2,000		3	765	Arrival at Lock 1	53	0	818 B chamber
5	17	3,200	8.5	3	785	Exit from Lock 1	761	0	1,546 At Lock 2
201	8	2,000	4.5	3	803	Exit from Lock 3	61	0	864 At Port 4
6	8	2,000	4.5	3	818	Exit from Lock 1	233	0	1,051 At Port 1
(1) 2	17	3,200		3	851	Arrival at Port 1	60	N/A	911
201	8	2,000		3	864	Arrival at Port 4	60	N/A	924
(1) 2	17	3,200	8.5	3	911	Exit from Port 1	494	0	1,405 At Lock 2
201	8	2,000	4.5	3	924	Exit from Port 4	737	0	1,661 At Port 3
202	17	3,200		3	945	Arrival at Lock 3	39	0	984 A chamber
202	17	3,200	8.5	3	984	Exit from Lock 3	1,239	0	2,223 At Lock 2
3	8	2,000		3	1,013	Arrival at Port 1	60	N/A	1,073
1	8	2,000		3	1,041	Arrival at Lock 2	35	0	1,076 B chamber

6	8	2,000	8.5	3	1,051	Arrival at Port 1	60	N/A	1,111	At Port 2
3	8	2,000	8.5	3	1,073	Exit from Port 1	154	0	1,227	At Port 3
1	8	2,000		3	1,076	Exit from Lock 2	68	0	1,144	B chamber
7	8	2,000	4.5	3	1,084	Arrival at Lock 1	44	0	1,128	At Port 2
6	8	2,000		3	1,111	Exit from Port 1	133	0	1,244	A chamber
8	8	2,000		3	1,121	Arrival at Lock 1	50	0	1,171	A chamber
203	17	3,200	5.5	3	1,124	Arrival at Lock 3	47	0	1,822	At Lock 2
7	8	2,000		3	1,128	Exit from Lock 1	694	N/A	1,204	
1	8	2,000		3	1,144	Arrival at Port 3	60	N/A	1,214	
4	17	3,200		3	1,154	Arrival at Port 2	60	0	1,722	At Lock 2
8	8	2,000	1.5	3	1,171	Exit from Lock 1	551	0	2,245	At Port 3
203	17	3,200	8.5	3	1,171	Exit from Lock 3	1,074	0	1,246	A chamber
9	17	3,200		3	1,180	Arrival at Lock 1	66	0	1,297	B chamber
10	17	3,200		3	1,188	Arrival at Lock 1	109	0	1,611	At Port 4
1	8	2,000	8.5	3	1,204	Exit from Port 3	407	0	1,556	At Lock 2
4	17	3,200	8.5	3	1,214	Exit from Port 2	342	N/A	1,287	
3	8	2,000		3	1,227	Arrival at Port 2	60	N/A	1,304	
6	8	2,000		3	1,244	Arrival at Port 2	60	0	2,007	At Lock 2
9	17	3,200	8.5	3	1,246	Exit from Lock 1	761	0	2,182	At Lock 1
3	8	2,000	8.5	3	1,287	Exit from Port 2	895	0	1,528	At Port 1
10	17	3,200	4.5	3	1,297	Exit from Lock 1	231	0	1,604	At Lock 2
6	8	2,000	4.5	3	1,304	Exit from Port 2	300	0	1,426	A chamber
204	8	2,000		3	1,355	Arrival at Lock 3	71	0	1,415	B chamber
205	8	2,000		3	1,369	Arrival at Lock 3	46	31	1,464	B chamber
206	8	2,000		3	1,386	Arrival at Lock 3	47	30	1,483	A chamber
207	17	3,200		3	1,400	Arrival at Lock 3	53	0	1,478	A chamber
(1) 2	17	3,200		3	1,405	Arrival at Lock 2	73			

FIGURE 14. Optional computer output.

N/A = Not available.

85

Total time: 44,000 minutes
Parameter arrival rates (mean number of tows per day): Down, 10.00; up, 10.00
Computed arrival rates: Down, 9.75; up, 9.88

Totals at	No. of tows	No. of barges	Chamber A — Time lost by tows passing through	Chamber A — Time chamber was processing a tow	Chamber B — Time lost by tows passing through	Chamber B — Time chamber was processing a tow	Total tow idle time at lock	Total lock processing time
			(· minutes ·)					
Lock 1	641	7,720	6,260	20,770	2,610	17,648	8,870	38,418
Lock 2	650	8,008	5,859	25,092	1,180	14,275	7,039	39,367
Lock 3	560	6,658	3,524	21,419	1,091	13,674	4,615	35,093

Delay statistics

Both chambers. Sub-values (in parentheses) are tows with delay of: Zero / LE10[a] / LE60[a] / GT60[a].

	Total delay time	Average delay time (minutes LE10[a])	Standard deviation (LE60[a])	Delay cost (dollars GT60[a])	Average queue length (tows)	Chamber A No. of tows	Chamber A No. of barges	Chamber B No. of tows	Chamber B No. of barges (No. of double lockings)
Lock 1									
Down	3,934 (215)	13.2 (4)	27.9 (59)	3,662 (20)	0.09	166	2,417	132	1,184 (14)
Up	4,936 (240)	14.4 (7)	27.3 (63)	4,873 (33)	0.11	186	2,784	157	1,337 (9)
Lock 2									
Down	3,636 (206)	11.9 (10)	22.6 (80)	3,685 (10)	0.08	168	2,586	138	1,212 (12)
Up	3,403 (244)	9.9 (10)	19.8 (85)	3,467 (5)	0.08	256	3,416	88	794 (10)
Lock 3									
Down	1,747 (201)	6.8 (11)	16.1 (40)	1,663 (6)	0.04	122	1,903	136	1,187 (11)
Up	2,868 (216)	9.5 (14)	19.2 (62)	2,879 (10)	0.07	230	2,947	72	621 (5)

Total system time (tow-minutes): 1,543,829. Total system cost: $1,454,321. Gross ton-miles: 1,983,503,814.
Lock delay time (minutes): 20,524. Delay cost: $20,228. Delay point time: 0. Cost: 0.

[a] LE = less than or equal to x minutes; GT = greater than x minutes.

FIGURE 15. Illustrative final computer output.

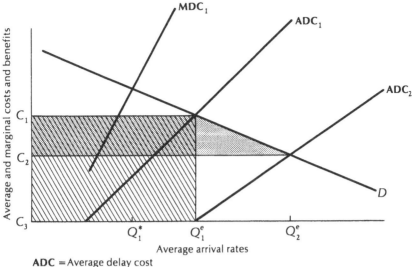

ADC = Average delay cost
MDC = Marginal delay cost

FIGURE 16.   Determination of optimum and equilibrium traffic flows.

erating costs. This leaves on the cost side only delay costs and charges for the use of the waterway. The optimum rate of utilization of the waterway is then determined by the rule:

Marginal delay (congestion) cost (MDC) = marginal willingness to pay net of straight-through operating costs.[7]

The solution is illustrated in Figure 16, in which Q_1^* represents the optimum traffic level. Only if all units using the waterway belong to one firm will the optimum equal the equilibrium traffic level, because a given firm will take into account only that part of the delay costs which it is required to bear. Under the competitive conditions obtaining in the industry, the equilibrium level will be Q_1^e, with the usual transport system inefficiencies of excessive delay costs. In the absence of administrative controls or a toll system that would impose the difference between average delay cost (ADC) and MDC on all users, Q_1^e would be the traffic level established.

Consider the impact of some system change (such as the installation of improved locks) on the structure of delay costs and the level of traffic. The threshold level of traffic beyond which delay costs occur will be shifted and the form of the ADC function may be changed, as with ADC_2 in Figure 16. If straight-through operating costs are changed by the system improvement, the demand curve, D, will also shift upward in such a way that the increased area under the curve at the old level of traffic will equal the total saving of straight-through costs at that volume of traffic. Since we are considering in this section the structure of congestion costs and since not all structural improvements need affect straight-through costs, the possible shift in the demand function is ignored here.

---

[7] Congestion phenomena have been analyzed in the literature of transportation economics for many years. For a particularly clear analysis in the context of highways, see Mohring [1962].

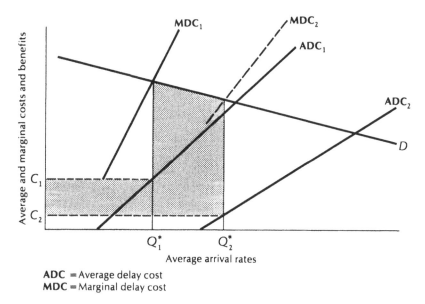

ADC = Average delay cost
MDC = Marginal delay cost

FIGURE 17.   Net benefits from waterway improvements under optimum traffic flow.

Assuming the absence of administrative controls or optimum tolls, a new equilibrium will be established at Q_2^e. The benefits from the system change can be broken into two components: (1) the cost savings on the previous level of traffic, $Q_1^e (c_1 - c_2)$; (2) the net benefits to the new traffic [the triangular portion of the shaded area; i.e., approximately $1/2(Q_2^e - Q_1^e)(c_1 - c_2)$].

If demand were completely inelastic, the latter component of benefits would be zero and total benefits would equal the delay cost savings on the unchanged level of traffic. Only in this case would cost savings on the previous level of traffic correctly measure benefits. If demand were completely elastic, both components would be reduced to zero as traffic increased to Q_2^e.[8]

To what extent can benefits attributable to system changes be measured in the absence of knowledge of the demand function? The difficulty is that they really cannot be correctly measured, and the measure which is available, namely the cost savings on the existing volume of traffic, may either understate or overstate total benefits, depending upon whether or not the optimum level of traffic is enforced. As illustrated in Figure 16, if the volume of traffic is held at Q_1^e, as it might be in running the simulation model to analyze delays before and after a system change, the measured cost savings would be $Q_1^e (c_1 - c_3)$, the cross-hatched area. The actual net benefits resulting from the induced change in traffic levels are illustrated by the shaded area. Thus, the measured cost savings on the unchanged level of traffic, Q_1^e, would *exceed* actual benefits.

Suppose, however, that some administrative controls or a toll system were in effect that could enforce the socially optimum level of traffic (see Figure 17). The initial traffic level would be Q_1^* and the level resulting from the waterway

---

[8] For demand to be completely elastic, a substitute mode of transport would have to be available at constant cost so that no rent (in the economic sense) would accrue to users of the waterway.

improvement would be Q_2^*. The true net benefits are illustrated by the shaded area of Figure 17, but the cost savings measured on the prior volume of traffic would be $Q_1^*(c_1 - c_2)$, an *understatement* of benefits.

It would be possible to assume alternative demand functions and to use them to estimate the new equilibrium level of traffic. This has not been done here, so the measure of net benefits remains that of the cost saving on a constant level of traffic.

Local Versus System Measures of Performance

In this chapter, empirically based measures of average and marginal delay costs are generated through computer simulation of a portion of an inland waterway transport network. Computer simulation, rather than an analytic queuing model, is primarily used to capture the contribution of all the elements of the network that jointly determine total system congestion cost; e.g., relations between delays at adjacent locks, relations between locking time and approach channel characteristics, and relations between locking time and tow characteristics. Recent experience on the Ohio River suggests the importance of fully incorporating these considerations into a systems analysis of inland waterway transport networks. When four high-lift dams in the busy central reaches of the river, covering some 300 river miles, were completed, "local" congestion in that area was greatly reduced. However, two events occurred which, if not unforeseen, were underestimated in their impact.

The first was that much of the existing delay time was apparently transferred downstream to Lock 52, which lies below the confluence of the Cumberland and Tennessee rivers with the Ohio River, is nine dams removed from the congested area, and has a more limited capacity. (See Figure 11.) The situation was aggravated by a second event: bargeline firms began to adapt the size of their barge flotillas to the larger (110- by 1,200-foot) lock chambers of the new dams, requiring double lockings at the remaining old (110- by 600-foot) locks. By the summer of 1966, these factors, in addition to the larger average traffic volume at Lock 52,[9] had caused the average time for passing through (delay and locking) Lock 52 to increase to the unusually long time of 3.5 hours.

The present simulation model does not incorporate any mechanism for forecasting adaptations of equipment to structural changes in the river system, but it does permit the analysis of all other impacts on the entire system of structural changes, including such delay transfers as have obviously occurred on the Ohio.

The behavior of average and marginal delay costs in a river system can be illustrated by reporting on simulation runs of a system consisting of three dams and two pools characteristic of the reach of the Ohio River from Meldahl Lock and Dam at river mile 435 to McAlpine Lock and Dam at river mile 610. (See

[9] Cargo passing through Lock 52 increased from approximately 16 million tons annually in 1955 to 27 million tons in 1964. Current levels of traffic on the central reaches of the Ohio range from 10 to 15 tows per day, but the level at Lock 52 has been approximately 20 tows per day.

TABLE 17.   Inputs, Outputs, and Delay Costs of a Three-Dam River System

Expected arrival rates per day		Total tows into system in 31 days (actual)	Total system delay cost over 31 days	Average delay cost per tow into system	Marginal delay cost per tow
Up	Down				
10	10	598	$22,900	$38	–
12	12	736	36,800	50	$101
14	14	854	68,000	80	265
16	16	987	118,900	120	382
18	18	1,092	217,400	199	938
20	20	1,214	587,600	484	3,035

Figure 11, p. 77) The system parameters used in these runs were informally derived, mostly on the basis of judgment. The costs were compiled over runs simulating approximately thirty days of experience. While it certainly cannot be claimed that the parameters used in these illustrative runs accurately reflect Ohio River conditions, the model is not grossly unlike the Ohio system. The summer average traffic flows on the central Ohio are in the range of ten to twelve per day in each direction, not including pleasure craft. Table 17 illustrates the dramatic increases in delay cost consequent upon rather modest increases in traffic.

Within the limitation of the benefit measure used here, we can ask: "To what extent will locally measured benefits from structural or operating rule improvements differ from system benefits?" Two illustrative situations will be given. The first shows the effects of a structural improvement in the center lock of a three-lock system that reduces mean time for the total locking operation at that lock by one-half (see Table 18). The second shows the effects of introducing a new and intuitively "better" set of priority rules at the center lock of this system. Although the measures used are minutes of delay time saved in a period of thirty-one days, the cost of an average tow used in this simulation is very close to one dollar per minute, so the savings figures can be read as dollars with little error.

When the arrival rates into the system are sixteen per day in each direction, the positive increment of system savings over the savings at Lock 2 suggest that system benefits may exceed local benefits at low traffic rates, although the difference may not be significant and may be due to random fluctuations in actual traffic levels between the simulation runs. Evidently there is a great overstate-

TABLE 18.   Local and System Savings in Delay Time Over Thirty-One Days After Structural Improvement of Lock 2 in a Three-Lock System

(minutes)

Arrival rates per day (up and down)	Apparent savings at Lock 2	Actual system savings
16	48,767	52,347
22	1,515,775	989,972

TABLE 19. Average Queue Lengths After Improvement of Lock 2 in a Three-Lock System (Arrival Rates = 22 per Day)

	Average queue length	
	Original	After change
Lock 1		
Up	5.37	10.13
Down	5.75	11.34
Lock 2		
Up	16.30	0.12
Down	18.43	0.16
Lock 3		
Up	0.88	1.62
Down	1.20	2.07

ment of benefits as measured at Lock 2 itself when arrival rates are raised to twenty-two per day. In this case, the discrepancy between local and system savings of nearly 526,000 minutes consists of *added* delays of 455,000 minutes at Lock 1 and of 71,000 minutes at Lock 3. The impact is also seen in the changes in average queue length shown in Table 19.

The second illustration shows system versus local delay time savings consequent upon a change in locking priority rules at Lock 2. The original rules were quite simple and reflected, with some added sophistication, the procedures currently used on the Ohio: i.e., first come, first served and the choice of lock chamber based on minimum *expected* total time for the locking operation.[10] The revised rules applied at Lock 2 are still first come, first served but the selection of chamber is loaded against swingarounds and double lockings by the addition of an artificial constant to the expected times for these operations. The results are shown in Table 20.

Two interesting features stand out in Table 20: (1) the new operating rules appear to be inefficient at low traffic levels; (2) the difference between local and system delay time savings changes sign with the level of traffic as it did in the preceding example. The new rules were based on the rationale that certain types of operations (double lockings and swingarounds), while beneficial to the tow currently being processed, tend to tie up the lock and delay other units unduly.[11] There can be no doubt that the new rules are more efficient, for system benefits are positive at all measured traffic levels. However, contrary to expecta-

[10] Locking priorities are described in U.S. Army, Corps of Engineers, *Regulations Prescribed by the Secretary of the Army for Ohio River, Mississippi River Above Cairo, Ill. and their Tributaries; Use, Administration and Navigation* (U.S. Government Printing Office, 1961). The usual rule is "first come, first served in alternating directions." In the Ohio River Division, when queue lengths exceed three on both sides of the lock, "three up, three down" is used. The Corps has studies under way to see if a more efficient order can be developed.

[11] In personal discussions, Alan Chandler of the Ohio River Division, Corps of Engineers, has indicated that when there are queues in both directions at Lock 52, the operating rule of taking three tows each way has been adopted because, for that lock, swingaround time is less than the difference between long and short entry time.

TABLE 20.   Local and System Delay Time Savings Over Thirty-One Days Consequent Upon Changed Locking Rules at Lock 2 in a Three-Lock System[a]

(minutes)

Arrival rates per day (up and down)	Local savings	System savings
10	−1,525	−1,177
16	9,051	17,170
18	35,546	46,080
20	132,504	40,688

[a] Negative figures indicate additional delay time.

tion, system benefits in this particular system reach a maximum at arrival rates of eighteen. Thereafter, local benefits no longer *understate* but *overstate* system benefits.

The two foregoing examples serve to illustrate the serious discrepancies that can exist between local and system benefits accruing from physical or operating rule changes to a system. The lesson to be learned is the necessity of looking at the larger system of which the local component (lock) is but a part. The three-lock system used to generate these illustrations is itself, of course, only a part of a still larger system. The question that remains is: How large a system must be included in the analysis if all significant impacts of a change are to be recognized? For example, the answer might be the entire Ohio River or the entire interconnected Mississippi system. This question will be addressed in the next section.

THE STRUCTURE OF CONGESTION COSTS AND OPTIMUM TRAFFIC FLOW

It has already been indicated that the optimum level of traffic on a waterway is that at which the marginal delay cost just equals the willingness to pay of the marginal user. It was also pointed out that the equilibrium rate of traffic (in the absence of an appropriate regulatory mechanism) would be established where the average delay cost equaled the marginal user's willingness to pay. Assuming that demand is high enough for average and marginal delay costs to be positive, the equilibrium traffic level would be inefficiently high, generating excessive amounts of delay. The equilibrium and optimum levels of traffic were illustrated by Figures 14 and 15.

Of course, various classes of traffic may have different marginal impacts on the performance of the system. Characteristics of traffic which would be expected, a priori, to lead to different impacts include: (1) size of the unit (e.g., number of barges); (2) draft of the unit (affects speed and maneuverability); (3) direction of travel (traffic may be imbalanced); and (4) itinerary followed (may lead back through bottleneck locks). If these different characteristics do cause different impacts on system performance (in particular, on delays within the system), then an optimum regulatory mechanism should reflect these differences. For example, if tolls were used, the optimum toll for a particular type of tow (holding the traffic level constant) should be the difference between the average delay cost experienced by tows of that class and the marginal delay costs they impose on the entire system.

This section reports on an investigation to determine to what extent different classes of traffic have different impacts on system delays. In the simulation runs used for this purpose, traffic was distinguished by tow *size* and *direction* of travel at time of entry into the system. The scheme was to hold the average arrival *rate* of downbound traffic constant while varying the upbound arrival rate. Since the upbound arrival rate refers to the aggregation of all classes of upbound traffic, it was necessary to vary simultaneously the probability distribution used to select the size of upbound units in such a way that the increments in the average upbound arrival rate would result in an equal incrementing of the *expected* upbound arrival rate of a particular size class of tows. In more formal terms, the scheme was as follows: let λ_u represent the average arrival rate parameter for upbound traffic. When an arrival occurs, the program randomly selects the size (number of barges and horsepower of boat) and draft of the tow according to two twelve-cell probability distributions whose probabilities may be listed as:

$$p_1^s, \ p_2^s, \ \ldots, \ p_i^s, \ \ldots, \ p_{12}^s$$

$$p_1^d, \ p_2^d, \ \ldots, \ p_i^d, \ \ldots, \ p_{12}^d$$

If it is desired to increment upbound traffic of size class i, the following simple adjustments to the above parameters are made:

1. $\lambda_u' = \lambda_u + 1$;

2. $p_i' = \dfrac{\lambda_u p_i + 1}{\lambda_u + 1}$;

3. $p_j' = \left(\dfrac{\lambda}{\lambda + 1}\right) p_i \ (i \neq j)$;

4. the p_j^d's remain unchanged.

These rules follow from the observation that the expected arrival rate of a particular size class i is $\lambda_u p_i^s$. It should be noted that incrementing traffic in this way changes the *mean* of the stochastic arrival pattern of a particular class of traffic. In any given run of the simulation, the actual number of arrivals naturally may differ from this mean value. In the particular simulation runs reported here, only two size classes were assigned positive probabilities: eight-barge tows pushed by 2,000-horsepower boats and seventeen-barge tows pushed by 3,200-horsepower boats.

Three general comments should be made at this point regarding the nature of the observations generated by the simulation. First, the observations incorporate sampling error just as would observations on the real system underlying the model. Thus, even if the mean value of a variable is changed between runs, the actual value of that variable (or some function of it) experienced in a particular run may change by a different amount or even in the opposite direction because of the "luck of the draw."

Second, statistics representing averages relating to some component of the system (e.g., a particular lock) will show a relatively larger variability than will the same statistics relating to the entire system (e.g., delay time for the system versus delay time for a particular lock).

Third, marginal variables, when measured directly from simulation observations, will tend to be quite erratic. Since all system variables have random components and since marginal variables (e.g., marginal delay costs) are composed of the ratio of differences of system variables, the random components of marginal variables are compounded of several sources of random variation. Since the random components of system variables as introduced in this program are mutually independent, differences between them will exhibit greater variation.

Table 21 through Table 24 present the simulation results and exhibit the magnitudes of average and marginal delay costs and the incidence of the system delay costs at the individual locks of the system. Three points stand out:

- At low levels of traffic (e.g., $\lambda_D = 10$, $\lambda_u = 10$) congestion costs are low. Current levels of traffic on the Ohio River appear to range from ten to fifteen per day, not counting pleasure craft.[12]
- Average delay costs per tow rise as traffic increases and become substantial at higher traffic levels.
- The difference between average and marginal delay costs attributable to a particular class of tows, indicating the magnitude of an optimum toll, becomes fairly large even at modest traffic levels.

Another interesting point arises from the comparison of delay costs and total costs as the arrival rates of small and large tows, respectively, are incremented. The increase in delay costs per tow and per million ton-miles is more rapid as the large tow arrival rate is increased. However, the increase in total cost per million ton-miles is less for the large tows than for the small. This is particularly noticeable when the comparisons are made for equal values of system ton-mile output rates. It will be noted that the system is producing ton-mile output at a rate of 3.1×10^9 per period when the total upbound arrival rate of twenty tows per days consists of fifteen small and five large tows, and that the rate is 3.9×10^9 per period when the upbound arrival rate consists of fifteen large and five small tows per day. Clearly, the operating economies of scale of the larger tows more than offset the greater imposition of system delays per tow.

The distribution of delays among the locks is indicated to some extent by the delay cost per tow processed for each lock as given in the last three columns of each of Tables 21 through 24. Given that it is the upbound rate which is being

---

[12] Except in the lower river around Lock 52 where traffic rates are approximately twenty per day.

TABLE 21. System Performance over a Period of Thirty-Five Days for Increments of Upbound Small Tows (Eight Barges, 2,000 Horse-power) with a Downbound Arrival Rate of Ten per Day[a]

Upbound arrival rate, small tows	Total system cost	Total delay cost	Total ton-miles produced	Total no. tows into system	Average total cost per 10^6 tm	Average delay cost per 10^6 tm	Average delay cost per tow into system	Marginal delay cost per tow into system	Average delay cost per tow locked		
									Lock 1	Lock 2	Lock 3
	(1,000 dollars)	(dollars)	(millions)		(·········· dollars ··········)				(·········· dollars ··········)		
5	1,656	25,134	2,319	681	710	11	37	–	14	13	9
7	1,842	31,061	2,472	761	750	13	41	74	13	14	12
9	2,029	39,474	2,564	842	790	15	47	104	11	20	13
11	2,122	37,734	2,647	883	800	14	43	–	14	16	12
13	2,324	48,254	2,787	974	830	17	50	116	13	19	16
15	2,524	65,159	3,113	1,033	810	21	63	287	14	21	26

[a] Upbound arrival rate for large tows = 5 per day.

TABLE 22. System Performance over a Period of Thirty-Five Days for Increments of Upbound Large Tows (Seventeen Barges, 3,200 Horsepower) with a Downbound Arrival Rate of Ten Per Day[a]

Upbound arrival rate, large tows	Total system cost	Total delay cost	Total ton-miles produced	Total no. tows into system	Average total cost per 10^6 tm	Average delay cost per 10^6 tm	Average delay cost per tow into system	Marginal delay cost per tow into system	Average delay cost per tow locked		
									Lock 1	Lock 2	Lock 3
	(1,000 dollars)	(dollars)	(millions)		(... dollars ....)						
5	1,656	25,134	2,319	681	710	11	37	–	14	13	9
7	1,918	40,820	2,616	757	730	16	54	206	20	16	16
9	2,207	58,563	3,019	851	730	19	69	189	19	26	21
11	2,309	61,650	3,036	883	760	20	70	96	22	22	23
13	2,560	92,966	3,409	958	750	27	97	418	21	36	35
15	2,932	156,465	3,862	1,046	820	41	150	722	22	49	66

[a] Upbound arrival rate for small tows = 5 per day.

TABLE 23. System Performance over a Period of Thirty-Five Days for Increments of Upbound Small Tows (Eight Barges, 2,000 Horsepower) with a Downbound Arrival Rate of Sixteen Per Day[a]

Upbound arrival rate, small tows	Total system cost	Total delay cost	Total ton-miles produced	Total no. tows into system	Average total cost per 10^6 tm	Average delay cost per 10^6 tm	Average delay cost per tow into system	Marginal delay cost per tow into system	Average delay cost per tow locked		
									Lock 1	Lock 2	Lock 3
	(1,000 dollars)	(dollars)	(millions)		(. . . . dollars . . . .)						
5	2,119	52,980	3,055	870	690	17	61	–	24	21	11
7	2,311	62,618	3,207	950	720	20	66	120	27	20	16
9	2,564	90,432	3,457	1,048	740	26	86	284	33	32	19
11	2,741	119,920	3,641	1,123	750	33	107	393	40	44	24
13	2,981	144,027	3,848	1,215	770	37	119	262	44	49	29
15	3,416	191,095	3,820	1,254	890	50	152	1,207	38	80	38

[a] Upbound arrival rate for large tows = 5 per day.

TABLE 24. System Performance over a Period of Thirty-Five Days for Increments of Upbound Large Tows (Seventeen barges, 3,200 Horsepower) with a Downbound Arrival Rate of Sixteen Per Day[a]

Upbound arrival rate, large tows	Total system cost	Total delay cost	Total ton-miles produced	Total no. tows into system	Average total cost per 10⁶ tm	Average delay cost per 10⁶ tm	Average delay cost per tow into system	Marginal delay cost per tow into system	Average delay cost per tow locked		
									Lock 1	Lock 2	Lock 3
	(1,000 dollars)	(dollars)	(millions)		($\ldots\ldots$ dollars $\ldots\ldots$)				($\ldots\ldots\ldots\ldots\ldots\ldots\ldots$)		
5	2,119	52,980	3,055	870	690	17	61	–	24	21	11
7	2,439	82,458	3,377	972	720	24	85	289	34	30	15
9	2,650	102,326	3,720	1,044	710	28	98	276	35	35	25
11	2,908	151,342	3,916	1,119	740	39	135	654	34	55	45
13	3,223	201,568	4,314	1,191	750	47	169	698	46	64	53
15	3,531	365,012	4,520	1,262	780	81	289	2,302	46	126	109

[a] Upbound arrival rate for small tows = 5 per day.

incremented, there appears to be a surprising degree of uniformity in average delay cost per tow processed at each lock. It should be stated both that the traffic rates under consideration lie well within the range of values for which this system has experimentally been determined to have finite steady state conditions, and that at higher rates certain locks are very likely to become bottlenecks.

The distribution of delays among the locks is perhaps better illustrated by Table 25, in which the time-average upbound and downbound queue lengths are shown for Tables 21–24.

TABLE 25.   Average Queue Lengths for Various Arrival Rates: Thirty-Five Days

Average arrival rate (tows per day)	Lock 1		Lock 2		Lock 3	
	Up	Down	Up	Down	Up	Down
$\lambda_D = 10$:						
λ_u (large) = 5						
λ_u (small) = 5	0.11	0.09	0.09	0.09	0.07	0.04
= 7	0.09	0.12	0.12	0.11	0.11	0.07
= 9	0.11	0.09	0.23	0.15	0.14	0.09
= 11	0.14	0.10	0.18	0.14	0.15	0.09
= 13	0.14	0.11	0.25	0.18	0.21	0.14
= 15	0.16	0.12	0.29	0.21	0.40	0.23
λ_u (small) = 5						
λ_u (large) = 5	0.11	0.09	0.09	0.09	0.07	0.04
= 7	0.17	0.14	0.16	0.11	0.14	0.09
= 9	0.16	0.17	0.28	0.20	0.22	0.13
= 11	0.22	0.17	0.25	0.17	0.25	0.17
= 13	0.24	0.14	0.47	0.27	0.46	0.25
= 15	0.28	0.14	0.73	0.39	0.98	0.52
$\lambda_D = 16$:						
λ_u (large) = 5						
λ_u (small) = 5	0.23	0.30	0.18	0.24	0.07	0.08
= 7	0.27	0.32	0.20	0.22	0.13	0.13
= 9	0.37	0.41	0.38	0.37	0.21	0.16
= 11	0.48	0.48	0.54	0.54	0.29	0.20
= 13	0.55	0.54	0.72	0.56	0.40	0.28
= 15	0.47	0.51	1.12	1.07	0.54	0.38
λ_u (small) = 5						
λ_u (large) = 5	0.23	0.30	0.18	0.24	0.07	0.08
= 7	0.37	0.40	0.31	0.33	0.14	0.11
= 9	0.40	0.42	0.41	0.37	0.25	0.20
= 11	0.45	0.34	0.75	0.56	0.56	0.37
= 13	0.62	0.54	0.97	0.70	0.74	0.45
= 15	0.67	0.51	2.03	1.41	1.63	1.03

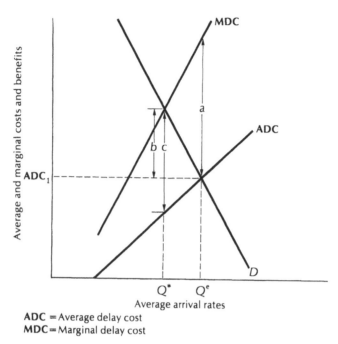

ADC = Average delay cost
MDC = Marginal delay cost

FIGURE 18.   Establishing the optimum traffic flow.

The desirability of some type of traffic regulation is made clear by the existence of uneconomic levels of congestion, even at fairly low volumes of traffic. Tolls have a clear advantage over all other forms of regulation in that they do not require waterway authorities to make judgments concerning who should use the river and who should not. Private decisions are made on the basis of the value to the tow operator of placing an additional tow on the waterway. Other regulatory schemes may restrict high-value traffic and favor low-value traffic.

However, some practical problems are involved in the setting of tolls. At present, no tolls are charged and traffic on some waterways may have grown to uneconomically high levels at which the difference between marginal and average delay costs is very large. The sudden imposition of high tolls would have a destabilizing impact on traffic, driving off the waterway not only the low-value traffic which is not economically justified but also traffic which should remain on the waterway at optimum traffic levels. This is illustrated in Figure 18. If tolls in the amount a were to be imposed on the traffic Q^e, the result would be to drive traffic to values below Q^*. Naturally, some adjustment in toll could be made as levels of traffic change but such changes would involve administrative work, would perhaps require congressional approval, and clearly would incur public displeasure. Since waterway users now perceive their average delay costs to be ADC_1, a toll in the amount b would raise the out-of-pocket costs of the users to the level of marginal delay costs at the optimum traffic level, Q^*. This cost would make unprofitable only the operations of those units whose presence on the river could not be justified at optimum traffic level, Q^*. As the level of traffic was adjusted downward toward Q^*, there would probably be a lag in users' perception of the changes in their delay costs. At some time before the average traffic level reached Q^*, the toll should be increased to c, again making the sum of average delay cost and toll equal to marginal delay cost. Thereafter, occasional increases in toll would be required if the demand curve shifted to the right over time.

Additional practical difficulties in the setting of optimum tolls are caused by the existence of several classes of traffic and by the lack of knowledge of the demand functions. As long as tolls could be changed experimentally (only a few changes might be required), these difficulties could be overcome. It will be remembered that the average and marginal delay costs can be estimated via simulation for any level of traffic and for each class of traffic. The following simple conditions then suffice to determine when the optimum toll and traffic level has been reached for each class:

- The level of traffic should have stabilized;
- At the stable traffic level, the average delay cost (known) plus the toll being charged must equal the marginal delay cost imposed by the particular class of traffic (also known).

It should be noted that, if the income distributional aspects of a toll system are objectionable in the eyes of Congress, rebates to bargeline firms could be made from the toll fund. Such rebates would not interfere with the attainment of an efficient traffic flow as long as they were not related to the levels of waterway use by the recipient firms. It is difficult, however, to conceive of rebate schemes that would be considered equitable that did not relate to the amount of waterway usage. The best use of the toll fund would thus probably be to finance public operation and maintenance expenses and, as far as funds permit, the construction of new facilities.

Additional Problems of Project Selection and Benefit Measurement

Two important and interrelated questions remain to be raised (if not answered):

1. If a waterway system must, for budgetary and operational reasons, be improved one project (e.g., one lock or one stretch of channel) at a time, how should the priority of projects be established? That is, how should the next project be identified?

2. Given that any improvement will have system impacts which extend beyond the local project area, how large a system must be encompassed in the analysis to account for all significant impacts?

The first question may be easy to answer in cases where the physical capacities or efficiencies differ substantially among existing locks and channels and where long queues are observed. Even here, however, it must be remem-

TABLE 26.   Average Delays Per Tow Locked, By Lock

Lock number	Delay in hours	
	Simulation run 1	Simulation run 2
1	6.2	8.8
2	5.7	15.9
3	6.4	5.6
4	13.0	12.4
5	9.4	7.8
6	4.2	7.6
7	13.5	3.4
8	5.5	5.5

bered that improving one facility alone may not be sufficient to alleviate a major part of the congestion. Where the locks of a system have approximately equal capacities (as will be the case on the Ohio River after completion of the present renovation program), the answer may not be at all obvious. In fact, there may be no "best" project to take on first. In particular, for a highly congested system of locks of equal physical capacities, observed differences in queue length may be due purely to random factors and may have no significance for the selection of a bottleneck facility for improvement. Two simulation runs were planned to determine how much variation in queue length might be expected among locks of identical characteristics in a larger (eight-lock) system. The eight locks were single-chambered (1,200- by 110-feet) and spaced thirty miles apart. Each was characterized by identical exponential locking times. Upbound and downbound arrival rates were set at ten and all tows passed straight through the system. These two runs of thirty-one days, under what turned out to be highly congested conditions, are summarized in Table 26. The point is that the lock with the greatest congestion appears to change from run to run even though all system parameters remain unchanged. Lock 7, which caused the greatest average delay in Run 1, had the lowest average delay in Run 2. Lock 2, which had the greatest average delay in Run 2, exhibited a low average delay in Run 1. The only consistency between runs appears at Lock 4 which has long delays in both runs. These observations suggest caution in attributing significance to queue lengths observed over short periods, especially within a system of uniform locks.

The second question cannot be answered with generality at this point, but two long (55.6 day) simulation runs with the model of Table 26 suggest that the relevant system may be fairly small, perhaps consisting of the next locks up- and downstream from the point of improvement. In the second run, the average locking times of locks 4 and 5 in the middle of the system were each reduced by 20 minutes. The results are exhibited in Table 27 in terms of average delays per tow locked at each lock. It appears that, in addition to a significant lowering of delay times at locks 4 and 5, there have been significant increases in delays at

TABLE 27. System Impacts of Lock Improvements

Lock number	Average delays per tow locked	
	Simulation run no. 1, all locks identical	Simulation run no. 2, locks 4 and 5 improved
	(......... minutes ....)	
1	92	80
2	65	64
3	60	76
4	57	14
5	78	12
6	60	77
7	63	57
8	63	66

locks 3 and 6. No other locks appear to have been affected. Of course, there is no way of applying a formal test of significance to the differences observed.

These results might be interpreted as suggesting that the relevant system for analyzing the system impacts of an improvement extends up- and downstream to the next lock and not beyond. This result warrants further investigation, however, both through analytical studies and simulation. The experience at Lock 52 on the Ohio River referred to earlier in this chapter indicated a transfer of delay time from the central reaches of the river where new locks and dams had replaced old ones to one of the old remaining locks (Lock 52) nine dams removed from the improved area.

CONCLUSIONS

It appears that simulation models can play an immediate and increasingly useful role in the management and planning of waterway systems. This methodology clearly permits a systems approach to management and public investment questions. The need for a systems approach has been made clear by the experimental runs of the present model, which indicated that the impacts of a particular system improvement may be felt in parts of the system far removed from the actual improvement. The calculation of system benefits is thus greatly complicated in the absence of a model of this type.

It was demonstrated for hypothetical cases similar to the modernized reaches of the Ohio River that congestion costs can be substantial at surprisingly low levels of traffic. It was also shown that tows of different sizes impose different incremental congestion costs on the waterway system. For example, in a three-lock system with a downbound arrival rate of ten tows per day (randomly assorted sizes), and an upbound arrival rate of five large tows (seventeen barges, 3,200 horsepower) and five small tows (eight barges, 2,000 horsepower) per day, average congestion cost per tow was $37. When the upbound arrival rate of small tows was increased to seven, the average cost rose to $41 per tow and the marginal cost per small tow was $74. If, however, the upbound arrival rate of large tows increased to seven, the average cost rose to $54 and the marginal cost per large tow was $206. These are, of course, average values over time, since the operation of the system is probabilistic in nature.

It was indicated that there can be a substantial difference between the average delay costs that are experienced by all tows and the marginal delay costs that are imposed by various sizes of tows. This difference represents the value of the economically efficient toll which could be imposed to bring about the most efficient level of traffic which, contrary to much common folklore, is *not* the greatest possible level of traffic.

Two important questions which remain largely unanswered were raised in this chapter. The first was: How should bottleneck points in a waterway system be identified in order to improve the system? It was demonstrated that, in larger systems in which the lock characteristics are similar, the apparent bottleneck (long queue) may shift about randomly over time, so that no particular significance could be given to the observation of queues in systems of that type.

The second question was: What should be the size of the system to be included in the analysis if it is to incorporate all of the effects of changes in the physical system? Very little information was generated on this point, although there was some indication that the required system might be smaller than expected.

APPENDIXES TO CHAPTER 5

Definitions

Component = constituent of the waterway system:
     permanent—locks, pools, ports, delay points;
     temporary—tows, pleasure craft.
  Attribute = property associated with a component, or with the system.
  Function = mathematical relationship describing the behavior of components.
    Event = that which creates or destroys a component; changes the numerical
     value of an attribute.

Attributes of components

	Inputs	*Outputs*
Tows	Number of barges	Total number of locks visited
	Horsepower	Total processing time
	Configuration	through locks
	Itinerary (ports visited,	Total delay time at locks and
	direction of travel)	delay points
	Arrival time in system	
	Draft	
	Tonnage	
Pleasure craft	Arrival time at a lock	
	Direction of travel	
Locks	Locking times	Number of tows processed
	(A lock, B lock; upstream	(upstream and downstream;
	and downstream; single,	A lock, B lock)
	setover, double, triple)	Number of barges processed
	Entry times	(upstream and downstream;
	(upstream and downstream;	A lock, B lock)
	long and short)	Total delay time
	Exit times	(upstream and downstream)
	(upstream and downstream)	Total cost of delay time
	Swingaround times	(upstream and downstream)
	(upstream and downstream)	Total of delay time squared
	Previous tow processed	(upstream and downstream)
	(its direction, completion	Distribution of delay times
	time, exit time)	(upstream and downstream)
		Total processing time
		(A and B locks)
		Number of double lockings
		Total delay time for pleasure
		boats
Pools	Channel depth	
	Channel width	
	Current speed	
	Length	
	(distance between locks,	
	ports, and delay points)	

Delay points	Channel depth	Number of tows processed
	Channel width	Total delay time
	Current speed	Total cost of delay time
	Length	Total of delay time squared
	Previous tow processed (direction and completion time)	Distributions of delay time
Ports	Distribution of processing time in port (constant in experimental runs)	
The waterway system	Present time	Total time for tows in the system
	Total time for simulation run	Total costs of total tow time
	Designation of tows in the system	Total lock delay time for tows
	Next event time	Total cost of lock delay for tows
	Location of next event	Total ton-miles of traffic moved
	Type of event	
	Direction of tow for event	

Functions

Speed function—relating tow speed to tow attributes and channel or delay attributes.
Cost function—relating tow operating costs to tow attributes and time in the system.
Locking function—selecting appropriate locking time distribution according to tow attributes.
Locking priority function—determining choice of chamber for a tow, priority for tows in a queue.
Delay priority function—determining priority for tows in a queue.

Events

Arrival of a tow in the system.
Departure of a tow from the system.
Arrival at a lock.
Arrival at a port.
Arrival at a delay point.
Exit from a delay point.

B. SYMBOLIC REPRESENTATION OF THE MODEL

Attributes of tows

$T*HP$	Horsepower
$T*NO$	Number of barges
$T*DF$	Draft (feet)
$T*LG$	Length (feet)
$T*WD$	Width (feet)
$T*AR$	Arrival time of tow at a point
$T*EX$	Exit time of tow from a point
$TL*OP$	Time to process tow through lock (including delay time)
$T*CS$	Operating cost per minute
$T*TN$	Tonnage
$T*DR$	Direction of tow

Attributes of locks

$L*SW$	Swingaround time
$L*SE$	Short entry time
$L*LE$	Long entry time
$L*LK$	Locking time
$L*EX$	Exit time
$L*CP$	Completion time of the locking operation through a chamber
$L*DY$	Lock delay time caused by congestion
$L*OP$	Operation time of the locking chamber
$L*CPJ$	Completion time of prior tow through a chamber
$L*EXJ$	Exit time of prior tow from the chamber
$L*DRJ$	Direction of prior tow using the chamber

Attributes of pools

$P*CU$	Current speed (miles per hour)
$P*DP$	Channel depth (feet)
$P*WD$	Channel width (feet)
$P*OP$	Operation time of tow using the pool
$P*SP$	Speed of tow in the pool
$P*DT$	Distance from point to point within the pool

Attributes of delays

$D*CU$	Current speed (miles per hour)
$D*DP$	Channel depth
$D*WD$	Channel width
$D*OP$	Operating time of tow using the delay
$D*DY$	Delay time of tow using the delay
$D*SP$	Speed of tow in the delay
$D*CP$	Completion time of tow processing through a delay
$D*DT$	Distance through the delay
$D*CPJ$	Completion time of prior tow through the delay
$D*DRJ$	Direction of prior tow through the delay

Attributes of ports

$PT*OP$	Processing time at the port

The locking operation

Case 1:	$T*DR = L*DRJ.$
If	$L*CPJ + L*SW - L*EXJ \leq T*AR;$
then	$L*CP = T*AR + L*SE + L*LK + L*EX$
	$L*DY = 0$
	$L*OP = L*SE + L*LK + L*EX + L*SW;$
otherwise	$L*CP = L*CPJ + L*SW - L*EXJ + L*SE + L*LK + L*EX$
	$L*DY = L*CPJ + L*SW - L*EXJ - T*AR$
	$L*OP = L*SE + L*LK + L*EX - L*EXJ + L*SW.$
Case 2:	$T*DR \neq L*DRJ.$
If	$L*CPJ \leq T*AR;$
then	$L*CP = T*AR + L*SE + L*LK + L*EX$
	$L*LD = 0$
	$L*OP = L*SE + L*LK + L*EX;$

otherwise
$$L^*CP = L^*CPJ + L^*LE + L^*LK + L^*EX$$
$$L^*DY = L^*CPJ + L^*LE - L^*SE - T^*AR$$
$$L^*OP = L^*LE + L^*LK + L^*EX.$$

In both cases:
$$L^*EXJ = L^*EX$$
$$L^*CPJ = L^*CP$$
$$L^*DRJ = T^*DR$$
$$T^*EX = L^*CP$$
$$TL^*OP = L^*CP - T^*AR.$$

Choice of chamber is based on a comparison of L^*CP for each chamber. The components in this calculation are expected values for the current operations (with perhaps penalties attached to certain operations). After the choice of chambers based on expected values has been made (minimizing TL^*OP), the calculation is repeated with randomly generated locking components. The selection of an appropriate locking time (L^*LK) is a function of tow size (T^*NO).

Operation in the pools

$$P^*SP = f(P^*CU, P^*DP, P^*WD, T^*LG, T^*WD, T^*DF, T^*HP, T^*DR)^{[13]}$$
$$P^*OP = 60 \cdot P^*DT/P^*SP$$
$$T^*AR = T^*EX + P^*OP.$$

Operation in the delay

$$D^*SP = f(D^*CU, D^*DP, D^*WD, T^*LG, T^*WD, T^*DF, T^*HP, T^*DR)$$
$$D^*OP = 60 \; D^*DT/D^*SP$$

Case 1: $T^*DR = D^*DRJ.$
  If $T^*AR + D^*OP \geq D^*CPJ;$
  then $D^*DY = 0$
               $D^*CP = T^*AR + D^*OP;$
  otherwise $D^*DY = D^*CPJ - (T^*AR + D^*OP)$
               $D^*CP = D^*CPJ.$

Case 2: $T^*DR \neq D^*DRJ$
  If $T^*AR \geq D^*CPJ;$
  then $D^*DY = 0$
               $D^*CP = T^*AR + D^*OP;$
  otherwise $D^*DY = D^*CPJ - T^*AR$
               $D^*CP = D^*CPJ + D^*OP.$

In both cases: $D^*DRJ = T^*DR$
               $D^*CPJ = D^*CP.$

Port operation

$PT^*OP = 60$ minutes.

Total time of tow in system

$$\sum_i PT^*OP + \sum_j TL^*OP + \sum_k P^*OP + \sum_m (D^*OP + D^*DY);$$

i = ports visited,
j = locks traversed,
k = pools traversed,
m = delays traversed.

[13] The exact form of the function and its rationale is reported on fully in Howe [1967].

Operating cost function

$$T^*CS = \frac{1}{1,440} \cdot [20 \cdot T^*NO + 550 + .246\ T^*HP - .00001(T^*HP)^2]$$

Tonnage function

$$T^*TN = \frac{1}{2,000} (52 + .44\ T^*DF)\ (T^*DF)\ (35)\ (195)\ (T^*NO)$$

Total ton-miles carried by tow in system

$$\sum_k (T^*TN)\ (P^*DT) + \sum_m (T^*TN)\ (D^*DT).$$

6

THE DEMAND FOR INLAND WATERWAY
TRANSPORTATION

In this chapter, an economic model describing the underlying structure of the demand for barge transportation in the Mississippi River System is constructed and tested. By this means a forecasting model is made available for the volume of traffic moved by barge over the principal routes of the Mississippi River System. This model will relate changes in the barge transport pattern to changes in regional industrial or agricultural activities and in the freight rates charged by the barge companies and their principal competitors, the railroads.

First, the demand for transportation between two spatially separated regions is derived graphically and algebraically from the regions' excess supply and excess demand equations. An analysis is then given of the demand for transportation when more than one mode is available. A graphical and mathematical analysis of an economic system containing more than two regions is given, including an analysis of the associated demands for transportation.

After the properties of the spatial equilibrium model are discussed, the model is simplified to a form in which it can be applied to the analysis of real data. This is done in two steps. First, the regional commodity exports and imports by barge are related by regression analysis to relevant economic variables in the various regions; i.e., barge and rail freight rates and measures of economic activity in each region. Then, these regional commodity exports and imports are considered as fixed and the model is reduced to a linear programming transportation model. The United States is divided into twelve regions, based upon the availability of barge commodity flow data and economic similarity of the regions. Instead of attempting to analyze all possible barge flows independently, the hypothesis that they constitute a least-cost trade pattern is used to predict these flows from the regional barge exports and imports.

The commodities considered are coal, grains, and iron and steel articles. The predictions of the individual flows based upon the model are then compared with the real flows. The model is found to provide a mechanism for relating changes in economic activity and freight rates to changes in the volume and pattern of barge traffic. A few hypothetical examples are calculated.

A SPATIAL EQUILIBRIUM FRAMEWORK

The purpose of this section is to demonstrate the derivation and analysis of the demand for transportation. The general spatial equilibrium model under analysis is as follows. For each of n regions for a given commodity there exists a commodity excess supply curve in terms of the local price in each region:

$$ES_i = f_i (p_i) .$$

Each region is separated from the others by a matrix of transportation costs: t_{ij} = unit transportation cost from region i to region j. The (positive) flow from region i to region j is represented by x_{ij}. The system is said to be in equilibrium when the following conditions hold:

$$\sum_{j=1}^{n} (x_{ij} - x_{ji}) \leq ES_i, \quad i = 1, \ldots, n;$$ (1)

$$p_j - p_i \leq t_{ij}; \text{if} <, x_{ij} = 0, \quad i, j = 1, \ldots, n.$$ (2)

Condition (2) implies that if $x_{ij} > 0$, then $p_j - p_i = t_{ij}$. The *demand for transportation*, d_{ij}, is defined as follows:

d_{ij} = the amount of the good that entrepreneurs in region i want to ship to region j, given the whole matrix of transportation costs.

It is clear that d_{ij} depends upon the shapes of the excess supply curves in each region and is thus a "derived" demand, like the demands for factors of production.

The Two-Region Spatial Equilibrium Model

The two-region model has been represented by the well-known "back-to-back" model frequently seen in connection with the theory of international trade. However, all the individual supply and demand curves need not be shown. In Figure 19, let ES_1 be Region 1's excess supply curve in terms of Re-

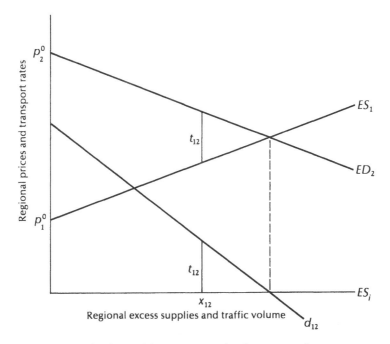

FIGURE 19. The demand for transportation in a two-region economy.

gion 1's price, and let ED_2 be Region 2's excess demand curve in terms of Region 2's price. Then, with no trade allowed between the two regions, the equilibrium prices in Regions 1 and 2 would be p_1^0 and p_2^0, respectively. However, with trade permitted at a constant unit transport cost t_{12}, entrepreneurs in Region 1 will want to ship an amount $ES_1 = x_{12}$ to Region 2. Plotting these quantities for shipment of each value of t_{12} gives the demand for transportation curve d_{12}. It is to be emphasized that d_{12} is in fact a demand curve. It gives the amount of a particular good that entrepreneurs in Region 1 want to ship to Region 2 for all "prices" of transportation. It can be seen that the only meaningful unit of measurement of this demand is a quantity of the particular good, per unit time, with the origin and destination specified. The same good flowing between two other localities is, in a very real sense, a different commodity, and any attempt to construct a "total" demand for transportation by mixing several regions and using a ton-mile measure introduces a degree of aggregation that makes the model structurally unsound. The actual value that t_{12} will take on will be determined by the intersection of the supply of transportation and the demand for transportation curves, or, in noncompetitive situations, by other appropriate equilibrium conditions.

Where the commodity excess supply and demand functions are linear, it can be shown that the demand for transportation function will also be linear and that its slope will be related in a simple way to the slopes of the excess supply and demand functions. In particular, if

$$ED_2 = a - bp_2$$
$$ES_1 = -c + dp_1 \; ;$$

then

$$d_{12} = \frac{ad - bc}{b + d} - \frac{bd}{b + d} \, t_{12}$$

$$= \frac{ad - bc}{b + d} - \frac{1}{\dfrac{1}{b} + \dfrac{1}{d}} \, t_{12} \, .$$

However, linearity of the trading regions' excess supply and demand functions is not a *necessary* condition for the linearity of the transport demand function.

The Demand for Transportation When Two Modes Are Available

The two-region model with two modes of transportation will now be explored. The direction of flow, say Region 1 to Region 2, is assumed known. Let:

$t_i =$ unit cost of transportation (to shippers) from Region 1 to Region 2, on mode i;

$q_i =$ flow on the ith mode per unit time ($i = 1, 2$);

$q_T = q_1 + q_2 =$ total flow per unit time;

$q_{Di}, q_{DT} =$ quantities of transport services on i^{th} mode demanded and the total demand, respectively;

$q_{Si}, q_{ST} =$ the corresponding supply of transportation services.

It is assumed that all qualitative differences between the two services (speed, risk, etc.) can be compensated for by an adjustment in price. The two services are thus regarded as perfect substitutes with respect to their adjusted prices. In Figure 20, in equilibrium, a total flow of q_T will be observed at a price t_T, and q_1 will flow on Mode 1, q_2 on Mode 2, where, by definition, $q_T = q_1 + q_2$. If Mode 2 decides to expand its operation on this route in a way that produces a rightward shift in its supply curve q_{S2} to q'_{S2}, an equal shift in q_{ST} to q'_{ST} takes place and more will be shipped at the lower price t'_T. Since $t'_T < t_T$, the amount carried by Mode 1 will decrease since q_{S1} is positively sloped. It can be noticed that Mode 2's expansion of service will be greater than that produced merely by the downward slope of q_{DT}, since it also takes business away from Mode 1. Since $q'_1 < q_1$, $t'_1 < t_1$, Mode 1 experiences a drop in revenue of the amount $t_1 q_1 - t'_1 q'_1$. In order to be able to say whether or not Mode 2 experiences an increase or decrease in total revenue, the separate or "partial" demands for the services of each mode must be explored further.

The demand for transportation was defined only in terms of moving a given commodity from one locality to another. This demand curve must then be independent of the mode of transportation. This being the case, it is not possible to speak of an absolutely defined demand for, say, barge transportation in terms of the price of this service. Thus, the demand curves $q_{Di}(t)$ are not simple demand curves like q_{DT}.

If Mode 1 represents barge transportation and Mode 2 refers to rail transportation, the demand curve for barge transportation must depend on the *supply* curve of the competitive mode, rail transportation [Marshall, 1948]. A shift

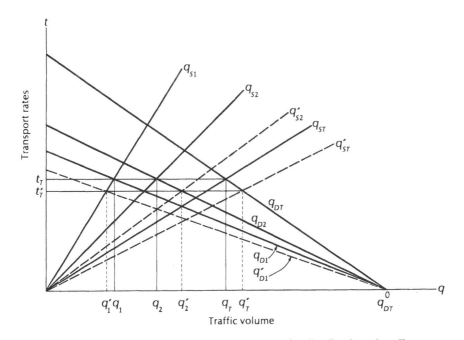

FIGURE 20. Effects of two modes of supply on the distribution of traffic.

in q_{S2} does not merely produce a movement along q_{D1}, it actually shifts q_{D1}. More precisely, the q_{Di} are defined as:

$$q_{D1}(t) = q_{DT}(t) - q_{S2}(t)$$
$$q_{D2}(t) = q_{DT}(t) - q_{S1}(t).$$

It can be noted from these equations and from Figure 20 that a shift in q_{S1} does not alter q_{D1}, so that the supply and demand curves for any one mode are in fact independent, as is necessary for a well defined system. Note that the shift in q_{S2} did not change the point of q^0_{DT}, the maximum value of transport demand at $t = 0$.

It can be shown that the demand and supply elasticities of the two modes are related in a simple but interesting way to the elasticity of total demand. In particular, the following relationship holds:

$$\left(\frac{q_1}{q_T}\right) E_{S1} + \left(\frac{q_2}{q_T}\right) E_{D2} = E_{DT} ;$$

where E stands for elasticity as indicated by the subscript.[1] This relationship may be rewritten as

$$E_{D2} = \frac{1}{\alpha_2} E_{DT} - \frac{(1 - \alpha_2)}{\alpha_2} E_{S1} ;$$

where α_2 is the *market share* of Mode 2. It follows that: (1) the elasticity of demand for Mode 2 becomes greater (in absolute value) as the total demand becomes more elastic and as the supply elasticity of the other mode increases; (2) as the market share of Mode 2 increases, the demand for Mode 2 becomes *more inelastic*; (3) as the market share of Mode 2 increases, the elasticity of demand for Mode 2 approaches the elasticity of total demand.

Several specific applications of this model follow. In case 1, Mode 2 (rail) is effectively price regulated (in the sense of a fixed price) at price $\bar{t}$, with a finite capacity $\bar{Q}$ to be offered. Figure 21 shows the individual and total supply curves of transportation and parts of various demand curves, where $\bar{t} < t_c$, t_c = price and $q_{S2}(t_c) = O\bar{Q}$. $q_{ST} = q_{S1} + q_{S2}$, and since q_{S2} is vertical for t above t_c, CD of q_{ST} is parallel to q_{S1}. In an unregulated system, $q_{ST} = OACD$. In a system where rails are regulated at $\bar{t}$, $q_{ST} = OMABCD$. At price $\bar{t}$, the total amount of transportation forthcoming would be $OR + O\bar{Q} = OS = \bar{t}B$. Now BC is parallel to q_{S1} since q_{S2} is (under regulation) vertical for all $t > \bar{t}$ and BCD is a straight line.

The effects of regulation of rails on the respective shares of the market captured by barge and rail may now be analyzed. Four different cases are distinguished below.

[1] See Appendix A, pp. 140-141, for derivation.

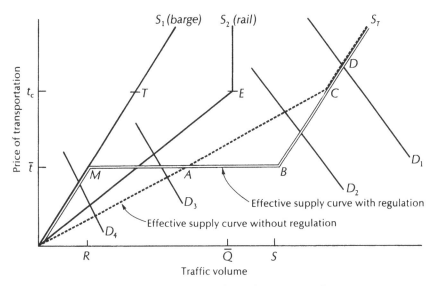

FIGURE 21. Effects of regulating rail transportation.

1. If q_{DT} cuts q_{ST} in the region CD, then regulation does not at all effect the absolute quantities that each mode ships or the price of barge service, since the aggregate supply curves coincide in this interval. $q_1 = q_{S1}(t \geq t_c)$, and $q_2 = O\bar{Q}$. However, rails are selling their service at $\bar{t}$ instead of $t \geq t_c > \bar{t}$.

2. If q_{DT} cuts through BC, rails sell $\bar{Q}$ at price $\bar{t}$, but now regulation has lowered both the price and the quantity of shipments by barge from what they would have been without regulation. That the price is lower is clear from the intersections of $q_{DT}(\text{II})$ with AC and BC, and q_1 is lower since this decrease in price lowers the *supply* of barges according to q_{S1}. If $q_{DT}(\text{II})$ cuts through AB, the only change in the analysis is that rails now ship less than $\bar{Q}$.

3. If q_{DT} cuts through MA, the equilibrium price of transportation will have been raised to $\bar{t}$ by regulation. With the higher price, more will be shipped by barge than otherwise would have been shipped in the absence of regulation, according to the Mode 1 supply curve. (One cannot tell by inspection whether shipments by rail have increased or decreased.)

4. If q_{DT} cuts through OM, then rails are cut out of the market entirely, and the amount shipped by barges is less than the total that would be shipped in the absence of regulation.

It may be noted that if the regulated price of rails were greater than or equal to t_c, q_{ST} would coincide with q_{S1} for $t < \bar{t}$. The reasoning of cases 1, 3, and 4 would apply. In conclusion, it has been seen that the effects of regulation depend greatly on the level of demand for transportation.

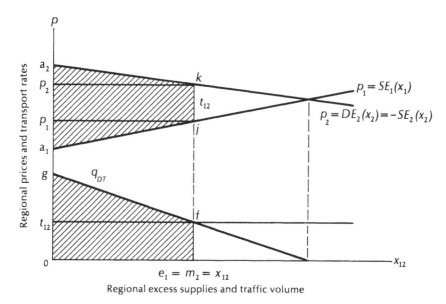

FIGURE 22.   Net benefits from trade, two regions, one commodity.

Maximization of Net Benefits in a Spatial Economy[2]

We begin with a two-region model of Samuelson, with the demand for transportation drawn in as in Figure 22. Let

$p_i = SE_i(x_i) =$ Region i's excess supply function expressed as a function of exported quantity.

$p_i = DE_i(x_i) = -SE_i(x_i)$ Region i's corresponding excess demand function.

$e_i =$ total net exports of Region i (possibly negative).

Let x_{ij} be the non-negative amount shipped from Region i to Region j; i.e., $x_{ij} \geq 0$. In terms of Figure 22, since all of Region 1's exports go to Region 2, and Region 2 does not export anything,

$$x_{12} = e_1 \geq 0 ,$$

$$x_{21} = 0 ,$$

$$x_2 = e_2 = -e_1 .$$

The reason we restrict $x_{ij} \geq 0$ will become apparent later.

Samuelson defines the "social payoff" in any region as the area under its excess demand curve, between zero and its net exports e_i. The value of the excess demand function $DE(x)$ for any quantity of output x is given by the value of the demand function less the value of the supply function, $D(x) - S(x)$. The inte-

---

[2] From Samuelson [1952].

gral of $DE(x)$ therefore represents the difference in areas under the demand and supply curves, measuring the net benefits from the given level of production, x. Then, net social payoff, NSP, is equal to social payoff in Region 1 plus social payoff in Region 2 minus total transport cost.

$$NSP = \int_0^{e_1} DE(x_1) \; dx_1 + \int_0^{e_2} DE_2(x_2) \; dx_2 - x_{12} \, t_{12} - x_{21} \, t_{21} \,. \qquad (3)$$

Now, in the two-region case of Figure 22, $e_1 = m_2$ and, by convention, $e_2 = -e_1$. Thus,[3]

$$NSP = -\int_0^{e_1} SE_1(x_1) \; dx_1 + \int_0^{e_1} SE_2(x_2) \; dx_2 - x_{12} \, t_{12} - x_{21} \, t_{21} \,. \qquad (4)$$

Expression (4) is exactly the shaded area between the excess demand and supply curves, $a_1 jka_2$ minus transport costs $0t_{12}fe_1$. Since the height of the transport demand function is always, by definition, equal to the vertical separation of SE_1 and DE_2, we must have $0gfm_2 = a_1 jka_2$. Thus NSP represents the area under the transport demand function less transport costs. This area is maximized when the marginal willingness to pay for transport is equated with t_{12}; i.e., when the gap between the excess supply and demand functions equals t_{12}. Maximizing NSP thus produces the equilibrium conditions (1) and (2) shown on p. 111.

The n-region form of Samuelson's NSP function is

$$NSP = -\sum_{i=1}^{n} \int_0^{e_i} SE_i(x_i) \; dx_i - \sum_{i,j=1}^{n} x_{ij} \, t_{ij} \,. \qquad (5)$$

Suppose that the e_i were given a priori, and that all that was left was to calculate the x_{ij}'s subject to the constraints of having e_i as total imports or exports in Region i. Then, since the equilibrium conditions are fulfilled only when NSP is a maximum, and since the integrals above are now constants, the equilibrium conditions will hold only when $\sum_{i,j=1}^{n} x_{ij} \, t_{ij}$ is as small as possible; i.e.,

when total transportation costs are minimized. Since this must hold for any set of e_i's it is clear that *any solution of the spatial equilibrium model must result in a minimum cost trade pattern for the particular e_i's prevailing in each region.*

This is an important result for the purposes of this study. It would be extremely difficult to estimate the excess-supply functions of all of the relevant regions of the United States for all commodities in order to derive the demands for transportation. Such estimation would require full structural knowledge of each region's supply and demand curves. The approach used in this paper takes advantage of the transportation cost minimizing property of optimum solutions to

---

[3] Notice that in the second integral, replacing DE_2 by $-DE_2 = SE_2$ changes the sign of the integral to negative, but changing e_2 to $-e_2 = e_1$ changes it back again.

the net benefit maximizing model. The e_i will be estimated for each region with the aid of various rail and barge rate data and measures of economic activity pertaining to the particular good involved. It is then relatively easy to derive the resulting flows from one region to another (the x_{ij}'s) by solving what is known as the linear programming transportation problem (LPTP). Actually only that part of the e_i that is shipped by *barge* will be investigated, and the resulting x_{ij}'s will pertain to flows by barge only. These flows of traffic by barge must still be minimum cost, given the exports and imports in each region, since *NSP* can still always be increased by reducing any component of total transportation cost that leaves the rest of the system unchanged.

THE LINEAR PROGRAMMING TRANSPORTATION PROBLEM

The essential definitions concerning the linear programming transportation problem (hereafter referred to as the LPTP) will now be developed. Let us conceive of a number of regions, mnemonically referred to as "factories," which have certain non-negative amounts of a commodity, a_i, available for export, where $i = 1, \ldots, m$. Let us then postulate a number of regions or "warehouses" where the good is demanded, letting the non-negative amount demanded by each warehouse equal b_j, $j = 1, \ldots, n$. Let c_{ij} be the (constant) unit cost of shipping the good from factory i to warehouse j, and let x_{ij} be the shipment from i to j, where $x_{ij} \geq 0$. The linear programming problem is then to find that shipping pattern which minimizes the total transportation cost, subject to the constraints that all supplies are exactly exhausted and demands are exactly fulfilled. Mathematically we write

$$\min \sum_{i}^{m} \sum_{j=1}^{n} c_{ij}x_{ij} = Z$$

$$\text{subject to } \sum_{j=1}^{n} x_{ij} = a_i \qquad\qquad i = 1, \ldots, m$$

$$\sum_{i=1}^{m} x_{ij} = b_j \qquad\qquad j = 1, \ldots, n.$$

$$x_{ij} \geq 0 \text{ all } i, j.$$

(6)

It appears that we have a linear programming problem with mn unknowns and $m + n$ constraint equations. However, one of these constraints is redundant. Thus we really have mn variables with $m + n - 1$ constraints. A basic solution for the system has $m + n - 1$ variables and a basic theorem of linear programming tells us that an optimum solution of the LPTP need contain no more than $(m + n - 1)$ $x_{ij} \geq 0$ and may have less.

Let us consider an example of four supplying regions and four demanding regions with the cost matrix (c_{ij}) and supplies (a_i) and demands (b_j) indicated below:

Supplying regions	Demanding regions				
	R_1^d	R_2^d	R_3^d	R_4^d	
R_1^s	1.0	2.0	3.0	5.0	$a_1 = 25$
R_2^s	3.0	1.0	2.5	3.5	$a_2 = 15$
R_3^s	3.5	2.5	2.0	2.5	$a_3 = 10$
R_4^s	4.5	3.0	3.5	1.0	$a_4 = 5$
	$b_1 = 4$	$b_2 = 16$	$b_3 = 15$	$b_4 = 20$	$\sum a_i = \sum b_j = 55$

The optimum solution to this problem can quickly be computed and is shown in the matrix below.

Supplying regions	Demanding regions				
	R_1^d	R_2^d	R_3^d	R_4^d	
R_1^s	4	6	15		25
R_2^s		10		5	15
R_3^s				10	10
R_4^s				5	5
	4	16	15	20	

Let us examine for a moment the optimal solution. Notice that even though the cheapest source as far as Region R_3^d is concerned is Region R_3^s, this source was not utilized at all for R_3^d. The explanation of this lies in the alternative uses of R_3^s's supply. If it in fact supplied R_3^d with the good, the other regions would have to make up their resulting deficit by shipments from other sources, which, in this case, would prove to outweigh the savings incurred by R_3^d. This suggests that it should be possible to place an implicit valuation on the various transport routes through an analysis of the dual problem of the LPTP. The variables of the dual are really indirect opportunity costs for the flows which do not appear in the solution.

THE EMPIRICAL MODEL

The waterway system modeled here consists of the Mississippi River from Minneapolis-St. Paul to the Gulf of Mexico; the Gulf Intracoastal Waterway from Brownsville, Texas, to Apalachee Bay, Florida; the Missouri River; the Ohio River, including the navigable portions of the Monongahela and Allegheny rivers; the Tennessee and Cumberland rivers; and some small tributaries of the Mississippi River System.

The definition of regions along the inland waterway system for purposes of recording import and export movements must be somewhat arbitrary. The first constraint on this selection is the form in which data are available. The only source of region-to-region barge shipment data is the Supplement to Part 5 of *Waterborne Commerce*, published annually by the Corps of Engineers [1956–63]. The origins and destinations are based upon the Corps of Engineers' Engineering Districts along the river system. No more detailed breakdown of regional data is available. Care should be taken not to aggregate diverse regions into one broad region, since the higher the degree of aggregation, the more the spatial qualities of the system will be lost. The twelve regions used are listed below with the corresponding engineering districts and the states which, for purposes of this study, have been assumed to be included in each district.

Region number	River district	State(s)
1	Gulf Intracoastal Canal, Brownsville, Texas to Louisiana border	Texas
2	All waterways in Louisiana below and including Baton Rouge	Louisiana
3	Gulf Intracoastal Canal, Louisiana-Mississippi border to Apalachee Bay Florida	Mississippi Alabama
4	Mississippi River, mouth of Ohio River to but not including Baton Rouge	Arkansas Tennessee (Mississippi)
5	Mississippi River, mouth of Missouri River to mouth of Ohio River	Missouri (Illinois)
6	Mississippi River, Minneapolis to mouth of Missouri River	Iowa Wisconsin Minnesota
7	Missouri River	Kansas Nebraska Missouri
8	Tennessee and Cumberland rivers	Tennessee (Kentucky)
9	Ohio River; Louisville, Kentucky, district	Kentucky Indiana
10	Ohio River; Huntington, West Virginia, district	Ohio West Virginia
11	Ohio River; Pittsburgh, Pennsylvania, district	Pennsylvania
12	Illinois River, port of Chicago, Illinois and Lake Michigan ports	Illinois

Certain compromises have to be made in defining the states associated with each region. The problem is that the waterways generally are on the boundaries between states while the other data in the model—the measures of industrial and agricultural activity and the rail freight rates—are published only on a state-by-state basis. Region 4, for example, touches both Arkansas and Tennessee, and also parts of Louisiana and Missouri. The ultimate assignment of states is based on the particular production and distribution patterns for each commodity. States for which data are used only for certain flows are indicated in parentheses. In particular, the assignments below are used.

For *coal*, Region 8, Tennessee and Kentucky are used, since most of the coal shipped on the Tennessee and Cumberland rivers originates in Kentucky.

For *grains*, Region 4, Mississippi is included in the analysis of import demands only. Since Region 5 touches much of Illinois, that state is included in the analysis of the supply of exports but is not included in the analysis of the demand for imports since grain shipments are mainly to St. Louis. Only Tennessee is used for Region 8 since most grains originate and terminate there rather than at other states through which the Tennessee and Cumberland rivers flow.

For *iron and steel articles*, Region 4 includes Mississippi and Region 8 includes Kentucky, since substantial shipments are made to and from locations in these states along the corresponding river regions.

The next step in implementing the model is the collection of data for the matrix of transportation rates by barge among the twelve regions, including the average rate incurred when a region makes a shipment to itself (i.e., to another port within the same region). Since the regions are not points in space but areas involving large distances within themselves, an average rate to and from each region must be constructed. This is done by using a major shipping or receiving city centrally located in each region, suggested by the data to be a reasonable approximation to a weighted center of each region. These major cities follow:

Region	City
1	Houston, Texas
2	New Orleans, Louisiana
3	Mobile, Alabama
4	Memphis, Tennessee
5	St. Louis, Missouri
6	Minneapolis, Minnesota
7	Kansas City, Missouri
8	Grand Rivers, Kentucky
9	Louisville, Kentucky
10	Huntington, West Virginia
11	Pittsburgh, Pennsylvania
12	(Peoria, Illinois)
	(Chicago, Illinois)

These central cities are occasionally varied to suit the needs of the model. For example, for *supply*, Peoria is used as the central port in Region 12, whereas *demand rates* are always calculated to Chicago. For Region 8, grain *demand* rates are calculated to Guntersville, Alabama.

The only source of barge freight rates is the barge industry itself. Rates can be either regulated or exempt from regulation. The relevant features of rate-making are as follows [Locklin, 1960]: Bulk commodities that do not require special handling when being loaded into a barge—e.g., coal, grains, scrap metal, sand and gravel, etc.—are exempt from regulation and are free to move at market rates if they are not mixed in barge flotillas containing more than three commodities or containing any regulated commodities. On the other hand, commodities that require special handling—e.g., iron and steel articles or other finished or semi-finished products—must move at the published rates. These rates can be changed only by approval of the Interstate Commerce Commission. Certain commodities, particularly coal and grain, although exempt under usual circumstances, generally move at or near the published tariffs, with occasional changes of rates. The rates for the three commodity groups used in this study were obtained from freight tariffs published by American Commercial Barge Lines [1956–63]. These tariffs are published jointly with the other leading firms in the barge industry and, with the exception of a relatively few smaller firms, represent the rates of the industry. Coal and grains are quite homogeneous from the standpoint of shipment by barge and rates are set only on a single commodity basis—i.e., coal, grains—ignoring any qualitative differences the commodities might have in their use. Various iron and steel items move at different rates. The rates on iron and steel articles given in the freight tariffs were used in this study.[4]

Two matrices of information are thus available: a matrix (x_{ij}) of observed flows from each region to all other regions by barge, computed from the Corps of Engineers' data, and the associated barge rate matrix (c_{ij}), for each of the eight years for which all data are available, 1956 through 1963. The row and column totals for the flow matrices are calculated for each (x_{ij}) matrix, the a_i's indicating row totals and the b_j's column totals.[5]

---

[4] The Waterways Freight Bureau collects complete data on actual rates charged by all bargelines on all shipments, but these data, collected in confidence, were unavailable. The opinion of industry authorities was, however, that the specific commodity rates did reflect the true rates of the industry.

Bargelines publish two sets of rates for *all* commodities, whether regulated or not: (1) "group" rates and (2) rates for "specific" commodities. For most major hauls, there is a specific rate that is never higher (and is usually much lower) than the group rate for that route. Theoretically, when a shipper wishes to ship a regulated commodity over a route for which there is no specific rate, the group rate will prevail. However, in practice it seems that the group rates are rarely used. If a specific rate has not been previously set, then for unregulated commodities (such as coal), a rate comparable to a specific rate will generally be arrived at by direct bargaining. The group rates thus represent hypothetical rates which might prevail in some unusual circumstances, and they are used in the cost matrix of this study only when no other rate is available. It can be observed, however, that no shipments over routes governed by group rates are predicted by the model, in conformity with the real world.

In constructing the shipping rate matrices, all applicable specific rates are entered first, and then the remaining elements are entered as group rates. The diagonal elements of the cost matrices (intraregional shipments) are estimates, since this type of data is not published. The construction of the rate for a region's shipments to itself is based on the time it takes to ship the commodity halfway across the region. This time is then compared with the times taken by shipments over similar routes for which data are available and the rates are assumed to be proportional to the time taken for the journey. The times taken for shipment used here are averages and were obtained from Federal Barge Lines, Inc.

[5] See Table 29, pp. 132-37.

The equations used for "prediction" of the a_i's and b_j's utilize the variables described below.[6]

Let Y_{it} = a particular a_i or b_j; i.e., total regional barge outflow or inflow of a particular commodity class, in tons, for Region i in year t, based on the flow data obtained from *Waterborne Commerce* [U.S. Army, Corps of Engineers, 1956–63];

I_{it} = measure of related economic activity in Region i and year t;

B_{it} = an indexed weighted average barge rate into or out of Region i in year t (see below for derivation);

R_{it} = an indexed weighted average rail rate into or out of Region i in year t (see below for derivation);

T_t = a time variable running from 1 to 8 in consecutive years (1956–63).

The functional form of the predicting equations that, in our judgment, was most accurate, was:

$$Y_{it} = \beta_0 + \beta_1 B_{it} + \beta_2 R_{it} + \beta_3 I_{it} + \beta_4 T_t .$$

In reality, shipments out of or into a region may depend on consumption and production in many regions as well as on competing rail and barge rates over many different routes, but only a limited number of observations was available because the Corps of Engineers did not start to collect region-to-region data until 1956.

The regression estimates of the import and export predicting equations are listed in Table 28 at the end of this chapter. Also listed are the mean value of the dependent variable ($\overline{Y}$) and the coefficient of determination (R^2) for each equation.

Table 29 presents an example of the model's application to the prediction of interregional coal flows for 1956, 1960, and 1963. The variables shown are the predictions of the complete model, $\overline{x}_{ij}$; the real coal flows, x_{ij}; predictions of coal flows based on actual totals of regional imports and exports and determined by cost minimizing, x'_{ij}; and the relevant barge shipping rates, c_{ij}. All flows listed in this table are expressed in *thousands of short tons* of coal, and all rate and total cost figures in the paper are expressed in *cents per ton*. All costs or flows not entered have zero value. A fictitious Region 13 is added when necessary to balance the constraints of the LPTP.

ESTIMATION PROBLEMS AND TESTS OF THE MODEL

The problems of formulating and estimating the regional equations have been difficult to solve. Admittedly, the variables which should enter these equations are not theoretically clear, and the interdependence of the rate index variables with the barge rates of the LPTP matrix is ignored by the model. Various estimating techniques were tried, but the estimates presented are single-equation least squares.

---

[6] Detailed definitions are given in Appendix B, pp. 141-42.

Some estimating problems have been imposed by the data characteristics that stem from the recent trends in the barge industry. Over the period of time covered by this study, barge rates have generally increased in response to increased demand for barge services and increased costs. At the same time, improvements in barge service and general economic expansion have led to continued gains in the volume of goods shipped by barge.

Simultaneously, the demand for rail transportation has been affected by increased competition from air freight, motor transportation, and bargelines. As a result of these competitive changes and technological improvements, rail rates in the years covered by the study have generally been falling. Thus, increasing barge flows have been accompanied by *increased* barge rates and *decreased* rail rates, making it difficult to identify the underlying import or export functions. The data points generally form *upward* sloping clusters when barge flows are plotted against the ratio B/R.[7] If these equations were interpreted as structural "barge export" and "barge import" equations, the regression coefficients of I_{it} and R_{it} should be positive, and the coefficient of B_{it} negative because higher rail rates, higher levels of economic activity, and lower barge rates would be associated with higher levels of commodity flows by barge. This is not always the case in these equations.

It will be noted in Table 28 that the time variable is often the most significant variable. This seems to indicate a trend associated with barge movements toward improved service, better materials handling systems, and a general increased awareness of the practicality of shipping by barge.

Aggregation. Basically, three types of aggregation are present in the model: aggregation of market areas into twelve large regions, aggregation of different qualities of goods into one supposedly homogeneous commodity, and aggregation over time of seasonal variations in demand and supply. Flows observed in the real data that are not predicted by the model probably can most frequently be attributed to the second type of aggregation. Examples of this are the actual coal flows to and from Pittsburgh (Region 11) and Huntington, West Virginia (Region 10). The data show that these two regions are cross-hauling coal: $x_{10,11}$ and $x_{11,10}$ are positive in each year, but $x_{11,10}$ is never predicted by the model. A reasonable explanation is that two products are involved: coal of a quality for coking, used in large amounts by the steel mills in the Pittsburgh area; and coal of a quality for conversion into electric power, used relatively more in the

---

[7] An example of this behavior follows. The data are for barge shipments of coal terminating on the Illinois River or in Chicago (Region 12).

Year	Y_t	B_t	R_t	I_t
1956	6,059	100.0	100.0	42,484
1957	6,215	100.3	101.2	42,718
1958	5,764	103.8	102.4	38,806
1959	7,316	104.8	98.6	39,720
1960	7,153	104.8	91.2	38,705
1961	7,453	109.8	91.1	37,479
1962	7,812	110.2	90.8	39,259
1963	7,520	110.2	94.4	39,086

Huntington district than in the Pittsburgh district. Unless the exact components of these two coal flows can be separated, the error caused by this type of commodity aggregation cannot be eliminated.

Aggregation of small regions into larger regions poses a less serious problem from the standpoint of consistent predictions. In fact, this type of aggregation can reduce prediction error although it may also reduce the usefulness of the prediction. In the previous example, if Regions 10 and 11 had been aggregated into one, then the cross-hauling between those two regions would have disappeared in the data. This type of aggregation is bound to affect the accuracy of the predicting equations since the variables in those equations will represent a greater aggregation of diverse regions than before.

Aggregation over seasonal variations can have the same effect as aggregation over different qualities of goods. The data are reported annually only and it is certainly possible for certain commodities to flow in different directions at different times of the year. This is not a serious problem in the case of coal. In the case of agricultural products it is more serious. For iron and steel articles, the data are highly aggregated from a quality standpoint and this is a more serious problem than seasonal aggregation with this commodity group.

Nonlinearity. The predictive barge import and export functions are not constrained by the model to be linear, but the assumption of constant unit transportation rates to the shipper is built into the model. In the short run, the nonlinearity of bargeline costs and hence, through competitive pressures, bargeline rates, is not likely to be important since most coal and grain rates are stable and iron and steel articles are regulated. In the longer term, economies of scale may affect barge rates. Some evidence on economies of scale was noted in Chapter 3.

ANALYSIS OF RESULTS

In order to get some quantitative idea of the accuracy of the model, one may observe the predicted and actual flows in Table 29. A summary measure of accuracy is a comparison between the total cost actually incurred by shippers and the total costs associated with the x_{ij}'s predicted by the model. In Table 30, three total cost figures are given for each of the three commodities. The costs that were actually incurred by shippers are shown in column (2). Column (3) shows what the costs would have been if the import and export equations gave completely accurate predictions and if the shipping patterns really minimized shipping cost. Thus the difference between columns (2) and (3) is a measure of the errors induced by aggregation (both by commodity and by region) which result in what then appear to be inefficient shipments. These errors are seen in column (5) to range from 5 per cent to nearly 10 per cent of actual costs. The costs actually predicted by the model are given in column (4), so that the differences between columns (4) and (2) represent the total cost error of the model. The differences between columns (4) and (3) measure the cost errors attributable to the export-import estimating equations.

The ratios of column (4) to column (2) can be taken as measures of the total relative error of the model. To adjust for the different tonnages involved in the predictions, (4)/(2) can be multiplied by (1)/(6), producing a ratio of costs per ton as predicted by the model *versus* the actual costs per ton. This avoids some offsetting quantity-cost errors and results in a relative error of between 7 per cent and 17 per cent.

A comparison of total costs may be too weak a test of the model since equality of predicted cost and actual cost does not imply that the predicted x_{ij}'s equal the actual x_{ij}'s.[8] Other measures of accuracy are given in Table 30. One manner in which to measure error in the model is to add the absolute value of differences between corresponding entries in any two of the three types of flow matrices. More precisely, the expression

$$\sum_{i,j} |\bar{x}_{ij} - x_{ij}|$$

gives the sum of absolute deviations between predictions and real flows. The lower limit on the value of this sum is clearly zero, which would occur if the predictions of the whole model exactly matched the real flows, while the upper limit is the sum of the total flows of the model and of the data; i.e., when all the "wrong" flows are predicted, the above sum would equal

$$\sum_{i,j} \bar{x}_{ij} + \sum_{i,j} x_{ij} .$$

The error can be partitioned into that associated with the predicting equations, and that due to aggregation and nonlinearity of the objective function. Using the identity:

$$\bar{x}_{ij} - x_{ij} \equiv \bar{x}_{ij} - x'_{ij} + x'_{ij} - x_{ij} ,$$

we get

$$|\bar{x}_{ij} - x_{ij}| \leq |\bar{x}_{ij} - x'_{ij}| + |x'_{ij} - x_{ij}|$$

and thus

$$\sum_{i,j} |\bar{x}_{ij} - x_{ij}| \leq \sum_{i,j} |\bar{x}_{ij} - x'_{ij}| + \sum_{i,j} |x'_{ij} - x_{ij}| .$$

These three sums are given in Table 31 in columns (1), (2) and (3), respectively.[9] The sums in column (2), since they compare two LPTP solutions, give the error caused by the differences between the estimated and real constraints; i.e., the error caused by the predicting equations. The sums in column (3) give a measure of error caused by the combined effects of aggregation and nonlinearity, as manifested in the use of our normative cost-minimizing hypothesis. The error of the whole model, and that caused by aggregation and nonlinearity, are then

---

[8] It might be noted that a cost structure (c_{ij}) which leads to an LPTP solution containing multiple solutions (non-unique optima) would imply cost indifference on the part of the shippers, making prediction quite difficult under any type of model.

[9] Since a "wrong" flow will appear twice, these sums are divided by two in the table.

given as a percentage of the total real flow. These percentages give a measure of error that appears to be larger than that given by the total cost comparisons.

APPLICATIONS OF THE MODEL AND CONCLUSIONS

The model can be applied to an analysis of the effects of factors that affect the demand for barge transportation. For example, assume that coal consumption in Region 11 (Pennsylvania) were to increase by 0.5 per cent from the 1963 level. Then, ceteris paribus, from the predicting equation for Region 11, total imports of coal by barge would be increased from 30,717 thousand tons to 30,870 thousand tons. The results of this particular change can be observed from Table 29(C). Since total demand has gone up by 153 thousand tons, the excess supply of Region 13 is increased by that amount. Proceeding on a path of flows predicted by the model and changing those flows by an amount that keeps the constraints satisfied, we see that $x_{10,11}$ must be decreased by 153; $x_{10,9}$ must be decreased by 153 to preserve row constraint 10; $x_{9,9}$ must be increased to preserve column constraint 9; $x_{9,1}$ must be decreased by 153; and $x_{13,1}$ must be increased by 153 to preserve row constraint 9 and column constraint 1, respectively. Thus the repercussions of this increased coal consumption throughout the rest of the system can be observed. The fictitious flow $x_{13,1}$, of course, means that Region 1 (Texas) does not receive its total demand of 309, but only receives $309 - x_{13,1}$. It should be noted that, in this model, the increase in consumption does not in itself tend to cause an increase of coal production in Pennsylvania; the increase in demand is satisfied by a shift in resources in the rest of the system, and tends to deprive Region 1 of adequate supply. This might then be expected to raise the price of coal in Region 1 to a level sufficiently high to induce more coal production, possibly from Pennsylvania. Thus some but not all of the spatial and general equilibrium characteristics of production, transportation, and consumption are observable with the use of the model.

In a similar manner, the effects of changes in barge rates can be observed. The procedure, however, is somewhat more hazardous because of the unstable nature of the transport pattern with respect to rate changes. Assume, for example, that one wishes to know the elasticity of demand for barge shipments from Illinois (Region 12) to the Tennessee River (Region 8). This flow is not predicted by the model in 1963 although a real flow of 843 thousand tons is present in the data. The rate $c_{12,8} = 300$ is just above the "indirect cost" $\bar{c}_{12,8} = 286$. It can be observed from the predicted solution that if $c_{12,8}$ falls by more than 14 cents/ton—i.e., by more than 4.76 per cent—a drastic rearrangement of shipments will be predicted; viz. Region 8 will now find it advantageous to receive grains from Region 12 rather than from Region 6 (the North Central States) causing $x_{12,8}$ and $x_{6,2}$ each to increase by 2,495 and $x_{12,2}$ and $x_{6,8}$ each to fall by 2,495, the latter flow becoming equal to zero. Thus the elasticity of this flow with respect to the corresponding barge rate will jump to infinity.

It is worth emphasizing that this instability is not solely a creature of this model. A perfectly competitive (or, for that matter, monopolistic) industry which is profit maximizing with constant unit shipping costs will tend to behave in this fashion. The transport pattern is not, of course, going to change overnight, owing to various institutional arrangements or other frictions. Given enough time, however, these large changes may come about in response to relatively small rate changes.

This model could be applied to analyzing the effects either of cost-saving waterway improvements or of putting various charges on the waterways. Certain assumptions would be necessary concerning the extent to which the bargeline firms would or could pass these savings or charges along to their customers. If the bargeline industry were highly competitive with each firm operating at the bottom of its long-run cost curve, cost savings or toll charges on a particular reach of the system would have to be entirely passed along to (absorbed by) the shippers (consumers of the barge services). Then the analysis could be carried out as before: by increasing the appropriate c_{ij}'s, and calculating the new b_i's and resulting changes in the barge flows. If, however, these savings or charges were partially absorbed by the bargelines, then additional information would be needed as to the change in the final rate charged to the shipper.

The potential benefit to be derived from models of this type is an ability to trace the impact on the entire barge transport system of changes in barge or rail rates or levels of activity in different regions. It naturally predicts "point-to-point" shipments in greater detail than aggregative models, but if one really wants to know how much corn will be shipped from the Upper Mississippi to the Tennessee region, a careful analysis of barge and rail costs, combined with surveys of potential sellers and buyers, will probably yield a more accurate answer than this type of model. The virtue of the model is its "systems" approach to determining mutually dependent demands.

Additional work is needed to structure the regional import-export equations in a more appropriate way. The evidence presented indicates, however, that this type of model can produce results of good accuracy and warrants additional development.

TABLE 28.   Regional Predicting Equations

Region	Barge outflow (1,000 short tons) $\overline{Y}$	Intercept β_0	Barge rate β_1	Rail rate β_2	Economic activity β_3	Time β_4	R^2
A. Coal exports by barge							
1	..	..	..	..	..	..	..
2	..	..	..	..	..	..	..
3	1,309	−427	0.166 (21.138)	1.451 (15.774)	0.048 (0.231)	2.081 (1.173)	.898
4	..	..	..	..	..	..	..
5	3,478	−2,234	−2.847 (15.498)	41.234 (47.954)	−0.170 (0.588)	5.662 (0.823)	.976
6	..	..	..	..	..	..	..
7	..	..	..	..	..	..	..
8	4,601	−3,954	9.816 (40.993)	−7.641 (9.688)	0.117 (0.031)	−1.407 (1.912)	.900
9	11,874	−97,339	1,032.270 (1,103.646)	84.161 (130.994)	−0.080 (0.047)	6.842 3.054	.965
10	18,052	−8,282	96.439 (45.119)	63.014 (33.459)	0.060 (0.017)	−1.961 (1.926)	.925
11	27,721	−49,712	11.277 (43.364)	302.800 (229.735)	0.571 (0.125)	8.129 (6.932)	.951
12	5,893	25,767	−146.443 (228.989)	−136.217 (76.665)	0.164 (0.160)	1.622 (5.337)	.796
B. Coal imports by barge							
1	193	1,969	−10.766 (13.048)	−9.106 (8.723)	0.054 (0.168)	0.531 (0.414)	.781
2	..	..	..	..	..	..	..
3	1,273	4,968	−7.480 (16.322)	−20.031 (14.401)	−0.112 (0.070)	2.840 (1.016)	.934
4	..	..	..	..	..	..	..
5	1,484	−3,433	−3.490 (3.423)	17.343 (18.059)	0.399 (0.158)	1.745 (0.462)	.944
6	3,156	474	26.923 (57.897)	−0.929 (3.597)	−0.032 (0.071)	2.381 (1.043)	.830
7	..	..	..	..	..	..	..
8	5,481	−9,764	78.383 (107.128)	−28.946 (80.022)	0.493 (0.107)	−7.870 (4.416)	.942
9	14,808	6,317	60.376 (103.598)	33.705 (114.384)	−0.043 (0.249)	3.350 (2.610)	.504
10	9,892	5,535	26.651 (17.449)	−51.594 (38.273)	0.110 (0.035)	−1.008 (0.892)	.970
11	29,511	914	5.888 (49.801)	−50.937 (214.852)	0.668 (0.211)	−0.227 (8.868)	.909
12	6,912	2,806	62.482 (174.650)	−77.598 (57.351)	0.113 (0.156)	1.112 (2.763)	.857

.. Not applicable.

TABLE 28.   Regional Predicting Equations—Continued

Region	Barge outflow (1,000 short tons) Y	Intercept β_0	Barge rate β_1	Rail rate β_2	Economic activity β_3	Time β_4	R^2
C. Grain exports by barge
1	..	..	..	..	..	..	..
2	..	..	..	..	..	..	..
3	..	..	..	..	..	..	..
4	..	..	..	..	..	..	..
5	1,097	−396	−4.434 (9.361)	3.172 (6.629)	0.001 (0.001)	0.210 (0.304)	.897
6	2,330	−3,118	5.476 (61.768)	25.912 (29.801)	0.000 (0.004)	4.530 (2.869)	.871
7	789	914	−3.473 (3.982)	−2.578 (1.796)	0.001 (0.000)	2.549 (0.450)	.986
8	34	−217	2.915 (5.448)	−0.426 (1.369)	0.000 (0.002)	0.141 (0.252)	.164
9	540	−2,796	42.179 (19.079)	−23.988 (11.707)	0.003 (0.001)	−0.773 (0.639)	.943
10	..	..	..	..	..	..	..
11	..	..	..	..	..	..	
12	3,169	−4,511	−15.220 (46.518)	24.197 (34.538)	0.006 (0.006)	3.661 (2.383)	.919
D. Grain imports by barge
1	143	−2,230	6.260 (2.647)	−1.324 (0.235)	3.461 (2.109)	1.676 (0.896)	.962
2	4,495	−1,961	17.370 (64.336)	−20.202 (26.308)	29.934 (29.012)	6.189 (3.534)	.962
3	384	−3,365	5.541 (4.246)	7.028 (5.437)	4.264 (2.927)	0.003 (0.856)	.943
4	532	−787	9.235 (10.524)	2.232 (3.879)	−0.473 (1.526)	1.102 (0.811)	.757
5	157	1,109	−5.072 (3.140)	−2.476 (4.642)	−0.211 (1.168)	−0.293 (0.271)	.637
6	61	−142	5.807 (2.510)	−2.558 (1.369)	−0.094 (0.074)	−0.131 (0.117)	.932
7	..	..	..	..	..	..	..
8	1,835	−4,205	122.479 (56.733)	−49.646 (24.599)	−9.561 (11.158)	4.000 (1.303)	.897
9	38	−234	1.351 (2.798)	−0.731 (2.084)	0.226 (0.360)	−0.014 (0.178)	.630
10	..	..	..	..	..	..	..
11	..	..	..	..	..	..	..
12	947	−35,897	135.809 (32.982)	−88.981 (32.745)	22.483 (6.244)	−18.457 (50.200)	.875

TABLE 28. Regional Predicting Equations—Continued

Region	Barge outflow (1,000 short tons) Y	Intercept β_0	Barge rate β_1	Rail rate β_2	Economic activity β_3	Time β_4	R^2
E. Iron and steel exports by barge							
1	164	299	−0.194 (1.646)	−0.539 (1.023)	−0.004 (0.552)	−0.118 (0.099)	.790
2	..	..	..	..	..	..	..
3	247	−2,833	31.995 (7.961)	−3.517 (1.498)	0.863 (0.515)	−1.130 (0.227)	.935
4	..	..	..	..	..	..	..
5	166	377	0.806 (5.769)	0.512 (0.994)	−1.432 (1.051)	−0.115 (0.127)	.794
6	63	354	−1.697 (5.409)	−0.301 (0.326)	−0.253 (0.207)	0.009 (0.117)	.688
7	..	..	..	..	..	..	..
8	..	..	..	..	..	..	..
9	104	203	−4.310 (6.592)	3.686 (3.127)	0.054 (0.099)	−0.115 (0.142)	.925
10	162	1,069	−8.794 (7.062)	−4.026 (1.143)	0.201 (0.060)	0.061 (0.120)	.907
11	3,133	−4,050	22.475 (45.725)	12.614 (9.794)	1.609 (0.408)	−1.833 (0.942)	.958
12	808	−1,330	31.603 (19.017)	−4.883 (12.836)	−0.604 (0.552)	−0.452 (0.509)	.722
F. Iron and steel imports by barge							
1	1,202	−2,610	56.414 (45.046)	−15.639 (9.267)	0.106 (0.393)	−2.834 (1.375)	.912
2	1,530	403	12.116 (10.583)	11.880 (11.161)	−0.911 (0.432)	−0.064 (0.675)	.888
3	87	−873	7.296 (5.631)	0.447 (1.604)	0.061 (0.090)	−0.381 (0.348)	.557
4	396	2,168	−11.946 (14.615)	−3.699 (1.475)	−0.048 (0.119)	0.288 (0.577)	.773
5	268	1,493	−10.791 (18.034)	0.252 (2.582)	−0.053 (0.262)	0.205 (0.693)	.395
6	182	−1,363	11.601 (11.983)	4.012 (3.915)	−0.002 (0.020)	−0.137 (0.321)	.827
7	123	−1,243	11.591 (11.399)	1.821 (0.954)	0.010 (0.163)	−0.064 (0.489)	.819
8	168	−417	2.159 (3.003)	2.240 (1.823)	0.048 (0.041)	−0.219 (0.109)	.804
9	668	2,360	−23.323 (9.590)	5.556 (3.090)	0.020 (0.047)	0.328 (0.376)	.902
10	69	−270	1.291 (2.445)	0.200 (0.434)	0.016 (0.009)	−0.156 (0.100)	.779
11	877	−961	11.471 (7.283)	−5.607 (2.774)	0.120 (0.047)	−0.619 (0.230)	.902
12	178	1,553	−14.805 (11.814)	2.891 (4.055)	−0.022 (0.044)	0.455 (0.318)	.757

TABLE 29. Shipments of Coal, 1956, 1960, 1963

(A) 1956

a. Full model prediction (*1,000 tons*).
b. Actual flow (*1,000 tons*).
c. Partial model prediction (*1,000 tons*).[a]
d. Cost per ton of transport by barge (*cents/ton*).[b]

		1	2	3	4	5	6	7	8	9	10	11	12	13	a_i
Region 1	a.	…	…	…	…	…	…	…	…	…	…	…	…	…	…
	b.	…	…	…	…	…	…	…	…	…	…	…	…	…	…
	c.	…	…	…	…	…	…	…	…	…	…	…	…	…	…
	d.	117	233	459	543	747	1,205	1,026	899	789	840	904	951	…	…
Region 2	a.	77	43	…	…	…	…	…	…	…	…	…	…	…	120
	b.	120	…	…	…	…	…	…	…	…	…	…	…	…	120
	c.	210	…	…	…	…	…	…	…	…	…	…	…	…	210
	d.	233	…	226	310	514	973	793	672	552	607	672	718	…	…
Region 3	a.	…	…	555	…	…	…	…	…	…	…	…	…	…	555
	b.	…	…	510	…	…	…	…	…	…	…	…	…	…	510
	c.	…	…	510	…	…	…	…	…	…	…	…	…	…	510
	d.	459	59	89	536	741	1,200	1,020	711	781	834	899	944	…	…
Region 4	a.	…	…	…	…	…	…	…	…	…	…	…	…	…	…
	b.	…	…	…	…	…	…	…	…	…	…	…	…	…	…
	c.	…	…	…	…	…	…	…	…	…	…	…	…	…	…
	d.	543	310	536	59	219	682	498	401	284	336	401	428	…	…
Region 5	a.	…	…	…	…	1,091	…	…	…	…	…	…	523	…	1,614
	b.	…	…	…	…	…	1,755	6	…	…	…	…	…	…	1,761
	c.	…	…	…	…	…	1,747	6	1	…	…	…	7	…	1,761
	d.	747	514	741	219	…	232	279	413	355	287	472	150	…	…
Region 6	a.	…	…	…	…	…	1,011	…	…	…	…	…	…	…	1,011
	b.	…	…	…	…	…	738	…	…	…	…	…	273	…	1,011
	c.	…	…	…	…	…	1,011	…	…	…	…	…	…	…	1,011
	d.	1,205	973	1,200	682	488	103	767	876	730	780	846	382	…	…

		1	2	3	4	5	6	7	8	9	10	11	12	13	Total
Region 7	a.	·	·	·	·	·	·	·	·	·	·	·	·	·	·
	b.	·	·	·	·	·	·	·	·	·	·	·	·	·	·
	c.	1,026	793	1,020	·	498	279	767	89	·	·	677	742	513	·
	d.	·	·	·	·	·	·	·	·	·	·	·	·	·	·
Region 8	a.	·	·	·	250	·	·	·	·	5,905	·	450	·	6	5,905
	b.	·	·	·	·	·	·	·	·	5,720	·	·	·	·	5,720
	c.	899	672	711	·	401	413	346	692	6,082	517	159	582	345	6,427
	d.	·	·	·	·	·	·	·	103	·	·	·	·	214	6,427
Region 9	a.	·	·	·	·	·	·	·	7,166	·	·	·	·	·	7,688
	b.	·	·	·	795	144	·	·	5,371	1,217	·	·	·	11	7,538
	c.	789	552	781	1,139	131	·	5,393	875	·	·	·	·	·	7,538
	d.	·	·	·	136	365	626	73	159	·	·	174	224	290	·
Region 10	a.	840	607	834	·	·	·	7,071	·	11,130	1,025	·	·	·	19,226
	b.	·	·	·	344	18	·	7,685	19	9,319	2,138	48	·	·	19,571
	c.	·	·	·	·	·	·	8,135	·	10,953	483	·	·	·	19,571
	d.	·	·	·	287	493	677	151	310	65	115	376	·	·	·
Region 11	a.	904	·	·	·	·	·	·	·	34,053	34,053	·	·	·	34,053
	b.	·	·	·	·	·	·	·	22	1,633	31,745	·	·	·	33,400
	c.	·	·	·	·	·	·	·	·	·	33,400	·	·	·	33,400
	d.	·	·	·	·	·	·	·	·	115	31	672	·	·	·
Region 12	a.	951	718	944	·	·	·	·	·	·	·	5,570	5,570	·	5,570
	b.	·	·	·	·	·	·	·	·	·	·	5,713	5,713	·	5,713
	c.	·	·	·	234	916	513	555	622	607	·	5,713	5,713	86	5,713
	d.	·	·	·	·	1,526	6	·	317	·	·	·	·	97	1,990
Region 13	a.	44	·	·	·	·	·	·	·	·	·	·	·	·	1,990
	b.	·	·	·	·	·	·	·	·	·	·	·	·	·	·
	c.	·	·	·	·	·	·	·	·	·	·	·	·	·	·
	d.	·	·	·	·	·	·	·	·	·	·	·	·	·	·
b_i	a.	121	598	1,091	2,537	6	6,744	14,237	11,130	35,078	6,190	·	·	·	77,732
	b.	120	510	1,139	2,897	6	6,957	13,528	10,953	33,883	6,058	·	·	·	76,051
	c.	120	510	1,139	2,897	6	6,957	13,528	10,953	33,883	6,058	·	·	·	76,051
	d.	·	·	·	·	·	·	·	·	·	·	·	·	·	·

· . Zero value or not applicable.

ᵃ The prediction given by the LPTP model but using actual regional imports and exports as constraint constants.

ᵇ The total barge transport cost for the predicted flow is $4,941,679; for the actual flows, $5,971,526; and for the partial model prediction, $5,544,719.

(B) 1960

a. Full model prediction (1,000 tons).
b. Actual flow (1,000 tons).
c. Partial model prediction (1,000 tons).[a]
d. Cost per ton of transport by barge (cents/ton).[b]

TABLE 29. (Continued)

		Region													
		1	2	3	4	5	6	7	8	9	10	11	12	13	a_i
Region 1	a.														
	b.														
	c.														
	d.	128	254	501	593	815	1,316	1,120	982	861	917	987	1,039		
Region 2	a.	37													37
	b.	37													37
	c.	37													37
	d.	254	64	246	339	562	1,063	866	734	606	662	734	784		
Region 3	a.			1,430											1,430
	b.	44		1,615											1,659
	c.	44		1,615											1,659
	d.	501	246	97	586	809	1,310	1,114	777	853	911	982	1,031		
Region 4	a.														
	b.														
	c.														
	d.	593	339	586	64	239	745	544	438	310	367	438	467		
Region 5	a.				824	334	2,720	4					538		3,970
	b.					1,466	2,182	4					401		4,053
	c.				824		2,426	4					799		4,053
	d.	815	562	809	153	24	182	305	451	388	282	516	144		
Region 6	a.						890								890
	b.						706						184		890
	c.						890								890
	d.	1,316	1,063	1,310	745	532	112	837	957	798	853	924	375		

	1	2	3	4	5	6	7	8	9	10	11	12	13	Total
Region 7 a.	..	..	..	..	..	..	..	..	..	..	..	..	..	..
b.	..	..	..	..	..	..	..	..	..	..	..	..	..	..
c.	..	..	..	..	..	..	..	..	..	..	..	..	..	..
d.	1,120	866	1,114	544	305	837	98	756	683	739	810	561	..	..
Region 8 a.	..	..	62	..	..	..	..	3,929	..	..	..	..	..	3,991
b.	..	..	..	..	..	75	..	3,855	..	..	..	32	..	3,962
c.	..	..	..	..	..	..	..	3,930	..	..	..	32	..	3,962
d.	982	734	777	438	451	311	756	112	164	565	..	209	..	..
Region 9 a.	205	626	..	..	1,399	..	..	1,024	9,105	..	..	..	..	12,359
b.	..	626	..	824	31	313	..	1,288	8,603	..	..	133	..	11,818
c.	179	582	..	..	1,725	..	..	1,264	8,068	..	..	..	..	11,818
d.	861	606	853	310	141	322	683	164	73	168	290	284	..	..
Region 10 a.	..	..	..	..	..	..	..	..	5,764	9,099	2,577	..	50	17,490
b.	135	..	..	..	228	40	..	51	5,979	7,573	3,577	78	..	17,661
c.	..	..	..	..	..	..	..	..	6,514	8,989	2,158	..	..	17,661
d.	917	662	911	367	282	498	739	310	112	71	134	376	..	..
Region 11 a.	..	..	..	..	..	..	..	..	..	..	25,345	..	..	25,345
b.	..	..	..	..	..	..	..	..	..	1,416	24,748	3	..	26,167
c.	..	..	..	..	..	..	..	..	..	..	26,167	..	..	26,167
d.	987	734	982	438	516	924	810	635	290	134	34	734	..	..
Region 12 a.	..	..	..	..	..	..	..	..	..	..	..	6,654	..	6,654
b.	..	..	..	..	..	..	..	..	..	..	..	6,321	..	6,321
c.	..	..	..	..	..	..	..	..	..	..	..	6,321	..	6,321
d.	1,039	784	1,031	467	756	1,004	561	679	606	662	734	91	..	..
Region 13 a.	..	..	..	..	..	..	..	..	..	..	..	..	..	..
b.	..	..	..	..	..	..	..	..	..	..	..	..	..	..
c.	..	..	..	..	..	..	..	..	..	..	..	..	..	..
d.	..	..	..	..	..	..	..	..	..	..	..	..	..	..
b_j a.	242	626	1,492	824	1,733	3,160	4	4,953	14,869	9,099	27,922	7,192	50	72,166
b.	216	626	1,615	824	1,725	3,316	4	5,194	14,582	8,989	28,325	7,152	..	72,568
c.	216	626	1,615	824	1,725	3,316	4	5,194	14,582	8,989	28,325	7,152	..	72,568
d.	..	..	..	..	..	..	..	..	..	..	..	..	..	..

.. Zero value or not applicable.

[a] The prediction given by the LPTP model but using actual regional imports and exports as constraint constants.

[b] The total barge transport cost for the predicted flow is $6,051,810, for the actual flows, $6,454,577 and for the partial model prediction, $6,075,429.

(C) 1963

a. Full model prediction (1,000 tons).
b. Actual flow (1,000 tons).[a]
c. Partial model prediction (1,000 tons).[a]
d. Cost per ton of transport by barge (cents/ton).[b]

TABLE 29. (Continued)

		Region													a_i
		1	2	3	4	5	6	7	8	9	10	11	12	13	
Region 1	a.														
	b.														
	c.														
	d.	148	274	521	613	835	1,336	1,140	1,002	881	937	1,007	1,059		
Region 2	a.														
	b.														
	c.														
	d.	274	84	266	359	382	1,083	886	754	626	682	754	804		
Region 3	a.		101	1,941											2,042
	b.	150	3	1,680											1,833
	c.		88	1,745											1,833
	d.	521	266	117	606	829	1,330	1,134	797	873	931	1,002	1,051		
Region 4	a.														
	b.														
	c.														
	d.	613	359	606	84	259	765	564	458	330	387	458	487		
Region 5	a.				1,167		3,367	39					435		5,008
	b.					1,730	2,837	7					189		4,763
	c.				1,167		3,378	39					179		4,763
	d.	835	582	829	153	44	190	325	471	408	282	536	149		
Region 6	a.						558								558
	b.						527								558
	c.						558								558
	d.	1,336	1,083	1,330	765	552	132	857	977	818	873	944	385		

		1	2	3	4	5	6	7	8	9	10	11	12	13	Total
Region 7	a.	..	..	..	..	..	..	..	..	..	..	..	..	..	..
	b.	..	..	..	..	..	..	..	..	..	..	..	..	..	..
	c.	..	..	..	..	..	..	..	..	..	..	..	..	..	..
	d.	1,140	..	..	1,134	564	325	857	118	776	703	759	830	581	..
Region 8	a.	886	..	..	..	..	..	..	4,753	..	..	..	500	..	5,253
	b.	..	..	..	..	..	29	..	5,194	..	..	..	115	..	5,338
	c.	..	..	..	..	..	..	..	4,671	..	..	..	667	..	5,338
	d.	1,002	..	..	797	456	471	311	132	164	585	655	214	..	..
Region 9	a.	184	891	2,042	1,166	..	472	32	..	10,084	..	12	465	..	14,341
	b.	810	..	9	..	..	..	..	..	11,062	..	..	..	..	14,316
	c.	303	904	1,902	..	..	..	..	..	10,137	..	..	..	..	14,316
	d.	881	626	141	330	..	330	703	..	73	168	..	289	..	..
Region 10	a.	153	..	..	..	..	163	..	..	5,636	10,290	2,598	..	..	18,524
	b.	..	178	..	..	..	..	..	..	4,047	8,630	4,725	..	..	18,337
	c.	..	..	..	..	..	..	..	..	5,003	10,395	2,939	..	..	18,337
	d.	..	..	..	..	..	47	..	498	112	91	134	69	..	..
Region 11	a.	..	..	..	..	..	..	..	..	..	..	28,119	..	..	28,119
	b.	..	..	..	..	..	..	3	..	31	1,753	26,644	..	..	28,431
	c.	..	..	..	..	..	..	..	..	..	..	28,431	..	..	28,431
	d.	..	..	..	..	..	..	..	54	310	134	..	..	376	..
Region 12	a.	..	..	..	..	..	..	..	..	..	..	..	6,723	6,723	6,723
	b.	..	1	..	..	..	21	..	..	..	..	..	6,652	6,652	6,674
	c.	..	..	..	..	..	..	..	..	..	..	..	6,674	6,674	6,674
	d.	1,059	..	..	..	..	..	..	..	626	682	54	..	96	..
Region 13	a.	125	..	..	..	..	..	..	581	699	754	754	96	96	125
	b.	..	..	..	..	..	..	..	..	..	..	..	..	..	..
	c.	..	..	..	..	..	..	..	..	..	..	..	..	..	..
	d.	..	..	..	..	..	..	..	..	..	..	..	..	..	..
b_j	a.	309	992	1,941	1,167	2,042	3,925	39	5,893	15,720	10,290	30,717	7,658	..	80,693
	b.	303	992	1,745	1,167	1,902	3,936	39	5,741	15,140	10,395	31,370	7,520	..	80,250
	c.	303	992	1,745	1,167	1,902	3,936	39	5,741	15,140	10,395	31,370	7,520	..	80,250
	d.	..	..	..	..	..	..	..	..	..	..	..	..	..	..

.. Zero value or not applicable.
a The prediction given by the LPTP model but using actual regional imports and exports as constraint constants.
b The total barge transport cost for the predicted flow is $7,968,258, for the actual flows, $8,497,133, and for the partial model prediction, $8,012,851.

TABLE 30. Performance of the Model: Costs and Flows

Year	Total real flows $\Sigma\Sigma x_{ij}$ (1,000 tons) (1)	Total real costs $\Sigma\Sigma c_{ij} x_{ij}$ (dollars) (2)	Total costs of predicted flows using actual constraints $\Sigma\Sigma c_{ij} x'_{ij}$ (dollars) (3)	Total costs of predicted flows using predicted constraints $\Sigma\Sigma c_{ij} \bar{x}_{ij}$ (dollars) (4)	Col. (3) as a percentage of col. (2) (5)	Total estimated flows $\Sigma\Sigma \bar{x}_{ij}$ (1,000 tons) (6)	Col. (4) ÷ col. (2) as a percentage of col. (1) ÷ col. (6) (7)
A. Coal							
1956	76,051	$5,971,526	$5,544,719	$4,941,679	92.85%	75,742	83.09%
1957	79,212	6,361,094	5,943,256	5,765,393	93.43	79,216	90.63
1958	66,082	5,600,942	5,146,928	5,182,909	91.89	67,708	90.32
1959	67,905	5,864,357	5,404,981	4,991,062	92.17	67,541	85.57
1960	72,568	6,454,577	6,075,429	6,051,810	94.13	72,166	94.29
1961	72,739	7,635,980	7,220,652	6,687,270	94.56	70,932	89.81
1962	75,314	8,078,156	7,654,794	7,486,585	94.80	76,405	91.35
1963	80,250	8,497,133	8,012,851	7,968,258	94.30	80,568	93.41
B. Grains							
1956	4,680	1,472,670	1,273,358	1,259,746	86.47	4,609	86.86
1957	4,248	1,749,556	1,496,863	1,406,161	85.57	4,810	70.98
1958	7,124	2,614,678	2,484,717	2,034,546	95.03	6,162	89.96
1959	7,926	3,014,333	2,804,273	2,846,874	93.03	8,133	92.03
1960	9,054	2,969,276	2,436,709	2,411,036	82.06	9,395	78.25
1961	9,283	3,161,832	3,043,645	3,261,948	96.26	9,707	98.66
1962	12,596	4,262,359	4,113,953	3,750,828	96.52	11,759	94.27
1963	13,937	5,028,419	4,836,946	4,778,411	96.19	14,045	94.30
C. Iron and steel articles							
1956	6,280	2,817,937	2,408,115	2,509,308	85.46	6,411	87.23
1957	7,174	3,437,079	3,069,524	2,852,059	89.31	6,919	86.04
1958	5,579	2,541,052	2,179,233	1,979,649	85.76	5,466	79.52
1959	5,633	2,603,983	2,084,895	2,048,952	80.07	5,781	76.68
1960	5,579	2,509,010	1,993,968	1,806,486	79.47	5,577	72.03
1961	5,259	2,327,530	2,000,951	2,009,027	85.97	5,579	87.37
1962	5,270	2,300,724	1,898,920	1,713,579	82.54	5,071	77.40
1963	5,211	2,269,634	1,877,291	1,792,125	82.71	5,153	79.85

TABLE 31.   Total Absolute Deviations (Coal, Grains, and Iron and Steel Articles)

(quantities in 1,000 tons)

Year	$\frac{1}{2}\Sigma\lvert\bar{x}_{ij} - x_{ij}\rvert$	$\frac{1}{2}\Sigma\lvert\bar{x}_{ij} - x'_{ij}\rvert$	$\frac{1}{2}\Sigma\lvert x'_{ij} - x_{ij}\rvert$	Col. (1) as a percentage of the total real flows	Col. (3) as a percentage of the total real flows
A. Coal	(1)	(2)	(3)	(4)	(5)
1956	9,145	4,729	5,007	12.02%	6.58%
1957	6,138	2,621	5,033	7.75	6.35
1958	8,952	3,301	7,171	13.55	10.85
1959	8,891	5,457	8,151	13.09	12.00
1960	6,136	2,555	7,012	8.46	9.66
1961	8,704.5	3,989.5	8,273	11.97	11.37
1962	10,252.5	2,706.5	9,596	13.61	12.74
1963	10,664.5	1,175.5	10,141	13.29	12.64
B. Grains					
1956	1,407	403	1,210	30.06	25.85
1957	1,943	1,046	1,672.5	45.74	39.37
1958	2,187.5	268.5	2,992.5	30.71	42.01
1959	3,149.5	712.5	2,980	39.74	37.60
1960	3,902.5	921.5	3,944.5	43.10	43.57
1961	3,549.5	1,481.5	3,123	38.24	33.64
1962	2,776.5	1,281.5	3,378	22.04	26.82
1963	6,228	896	5,470	444.69	39.25
C. Iron and steel articles					
1956	3,074	265.5	3,096	48.95	49.30
1957	3,507.5	345	3,505.5	48.89	48.86
1958	3,090.5	680.5	2,983	55.40	53.47
1959	3,031	568.5	3,131	53.81	55.58
1960	2,881	541	3,008	51.64	53.92
1961	3,113	449	2,816.5	59.19	50.67
1962	2,215.5	681.5	2,528.5	42.04	47.98
1963	2,477	299	2,488	47.53	47.75

$\bar{x}_{ij} \equiv$ the flow from region i to region j that is predicted by the LPTP using the constraints estimated with the predicting equations;

$x'_{ij} \equiv$ the flow predicted by the LPTP using the actual regional imports and exports as constraint constants;

$x_{ij} \equiv$ the real flows.

APPENDIXES TO CHAPTER 6

Let TR_i = revenue of ith mode. Then

$$q_{D2} = q_{DT} - q_{S1}. \tag{1}$$

$$TR_2 = tq_{D2} = tq_{DT} - tq_{S1}. \tag{2}$$

It follows that

$$\frac{dTR_2}{dt} = t\frac{dq_{D2}}{dt} + q_{D2} = t\frac{d_{DT}}{dt} + q_{DT} - t\frac{dq_{S1}}{dt} - q_{S1} \tag{3}$$

or,

$$\frac{dTR_2}{dt} = q_{D2}(1 + E_{D2}) = q_{DT}(1 + E_{DT}) - q_{S1}(1 + E_{S1}) \tag{4}$$

$$\frac{dTR_2}{dt} \gtrless 0 \text{ when } q_{DT} + q_{DT}E_{DT} - q_{S1} - q_{S1}E_{S1} \gtrless 0. \tag{5}$$

However, in equilibrium,

$$q_{Si} = q_{Di} = q_i. \tag{6}$$

Since $\qquad q_T = q_1 + q_2,$

$$\frac{dTR_2}{dt} \gtrless 0 \text{ when } q_1 E_{DT} + q_2 E_{DT} - q_1 E_{S1} + q_2 \gtrless 0, \tag{7}$$

$$\frac{dTR_2}{dt} \gtrless 0 \text{ when } q_1(E_{DT} - E_{S1}) + q_2(E_{DT} + 1) \gtrless 0. \tag{8}$$

We note also from equation (4) that

$$q_{S1}(1 + E_{S1}) + q_{D2}(1 + E_{D2}) = q_{DT}(1 + E_{q_{DT}}) \tag{9}$$

or in equilibrium, more generally,

$$q_1(1 + E_{S1}) + q_2(1 + E_{D2}) = q_T(1 + E_{DT}) \tag{10}$$

and expanding and using equation (1)

$$q_1 E_{S1} + q_2 E_{D2} = q_T E_{DT} \tag{11}$$

or

$$\left(\frac{q_1}{q_T}\right) E_{S1} + \left(\frac{q_2}{q_T}\right) E_{D2} = E_{DT}. \tag{12}$$

These last two equations give the elasticity of Mode 2's demand curve in terms of the elasticity of total demand, the elasticity of supply of Mode 1 and the proportions of total traffic carried on each mode.

Equation (12) may be rewritten as

$$E_{D2} = \left(\frac{q_T}{q_2}\right) E_{DT} - \left(\frac{q_1}{q_2}\right) E_{S1}. \tag{13}$$

140

As an example, if $E_{DT} = -0.3$, $E_{S1} = 0.5$, $q_1/q_T = 0.4$, and $q_2/q_T = 0.6$, then $E_{D2} = -0.83$. From (4) it is seen that $dTR_2/dt = 0.17\ q_{D2} > 0$.

It is clear from equation (13) that the demand for Mode 2 becomes more elastic as the elasticity of total demand increases and as the supply elasticity of the other mode increases. Further, equation (13) can be rewritten as

$$E_{D2} = \left(\frac{1}{\alpha_2}\right) E_{DT} - \frac{(1 - \alpha_2)}{\alpha_2}\, E_{S1}\ ; \tag{14}$$

where $\alpha_2 = q_2/q_T$. Differentiation of (14) with respect to α_2 yields:

$$\frac{dE_{D2}}{d\alpha_2} = \frac{1}{\alpha_2^2}\,(-E_{DT} + E_{S1}) > 0 \tag{15}$$

indicating that as the market share of Mode 2 increases, its demand becomes more *inelastic*, approaching E_{DT} as shown by equation (14).

B. DETAILED DEFINITIONS OF VARIABLES USED IN REGIONAL IMPORT AND EXPORT EQUATIONS

The variables, in detail, are as follows:

1. I_{it}. For the commodity "coal," this variable is defined as total regional coal production or consumption for the barge export ("supply") and for the barge import ("demand") equations, respectively. These data were obtained from the *Minerals Yearbook* [U.S. Bureau of Mines, 1956–63]. Consumption data for 1956 are estimates.

For the commodity group "grains," I_{it} is grain production (corn, wheat, and soybeans) or value added by manufacture in food and kindred products, for the supply and demand equations respectively. These data were obtained from *Agricultural Statistics* [1956–63] and *The Annual Survey of Manufacturers* [1956–63]. For the Region 2 (Louisiana) demand function, I_{it} is an index of grain exports from *Agricultural Statistics*, since most grains shipped to Louisiana are for export.

For the commodity group "iron and steel articles," I_{it} is value added by manufacture in primary metal industries, and total value added by manufacture for the supply and demand equations respectively. These data were obtained from *The Annual Survey of Manufacturers* [1956–63].

2. B_{it}. This variable is constructed in two steps: First, the barge rate matrices (c_{ij}) are indexed by setting all 1956 rates equal to 100 and determining the index for later years by dividing the year's c_{ij} by the corresponding c_{ij} in 1956 and multiplying by 100. Call this $\hat{c}_{ij}(t)$. Second, using these indexed rates, the total "revenue"

$$\sum_j \hat{c}_{ij}(t)\ x_{ij}(t) \text{ or } \sum_i \hat{c}_{ij}(t)\ x_{ij}(t)$$

for all shipments into or out of each region (for demand and supply) is calculated on the basis of the (x_{ij}) matrix and the $(\hat{c}_{ij})$ matrix. This total figure is divided by total barge imports or exports, giving an indexed weighted average barge rate per ton into or out of each region.

The indexing of the rates is an attempt to remove a troublesome feature of an unindexed rate. The shipping pattern is very sensitive to the rate structure over the whole system. With unindexed rates, if, for example, certain rates are lowered, the corresponding change in the trade pattern may be in the direction of longer average hauls, thereby *in-*

creasing the weighted average rate. Also, if a decrease in I_{it} tended to eliminate longer shipments before short shipments, producing a lower average rate, I_{it} and R_{it} would not be independent. In short, R_{it} would not be an exogenous variable, being dependent on Y_{it} and possibly on I_{it}. By adjusting the rates to the same base level, the effects of changes in length of haul are minimized.

3. R_{it}. This variable is constructed in the same manner as B_{it}, based on the Interstate Commerce Commission's one per cent waybill sample data of commodity movements by rail [U.S. Interstate Commerce Commission, 1956–63].

Rates per ton rather than per ton miles are used in the model.

BIBLIOGRAPHY

American Commercial Barge Lines, Inc. *Freight Tariff 3-B* and *Supplements*. Jeffersonville, Indiana, 1956–63.

Carroll, Joseph L. *Capacity of a Waterway Lock Via Simulation*. University Park: Pennsylvania State University, College of Business Administration, Center for Research, 1968.

Eckstein, Otto. *Water-Resource Development: The Economics of Project Evaluation*. Cambridge, Mass.: Harvard University Press, 1961.

Harris, Carl M. "Queues with State Dependent Service Rate," *Operations Research*, January–February 1967.

Hitch, Charles J. "Program Budgeting," *Datamation*, September 1967.

Horton, Clarence R., Jr. "Scientific Developments in River Transportation," *Journal of the Waterways and Harbors Division, Proceedings of the American Society of Civil Engineers*, September 1958.

Howe, Charles W. "Process and Production Functions for Inland Waterway Transportation." (Institute Paper No. 65, Institute for Quantitative Research in Economics and Management, Krannert Graduate School, Purdue University.) January 1964.

———. "Models of a Bargeline: An Analysis of Returns to Scale in Inland Waterway Transportation." (Institute Paper No. 77, Institute for Quantitative Research in Economics and Management, Krannert Graduate School, Purdue University.) July 1964.

———. "Mathematical Model of Barge Tow Performance," *Journal of the Waterways and Harbors Division, Proceedings of the American Society of Civil Engineers*, November 1967.

———. "Methods for Equipment Selection and Benefit Evaluation in Inland Waterway Transportation," *Water Resources Research*, Vol. 1, No. 1, 1965.

Inter-Agency Committee on Water Resources. *Proposed Practices for Economic Analysis of River Basin Projects*, Report by the Subcommittee on Evaluation Standards. Washington, D.C.: Government Printing Office, May 1950, revised May 1958.

Kelso, C. Edwin. "Factors Affecting Towboat Power Requirements." Master's thesis, University of Tennessee, August 1960.

Kneese, Allen V. *The Economics of Regional Water Quality Management*. Baltimore: The Johns Hopkins Press, for Resources for the Future, Inc., 1964.

Krutilla, John V., and Otto Eckstein. *Multiple Purpose River Development: Studies in Applied Economic Analysis*. Baltimore: The Johns Hopkins Press, for Resources for the Future, Inc., 1958.

Lave, Lester B., and Joseph S. DeSalvo. "Congestion, Tolls, and the Economic Capacity of a Waterway," *Journal of Political Economy*, June 1968.

Leininger, William J. "An Empirical Production Function for Barge Towing Operations on the Ohio River." Ph.D. thesis, Purdue University, August 1963.

Locklin, D. *Economics of Transportation*. Homewood, Illinois: Richard D. Irwin, Inc., 1960.

Maass, Arthur, *et al. Design of Water Resource Systems*. Cambridge, Mass.: Harvard University Press, 1962.

Marshall, A. *Principles of Economics*. 8th edition. New York: Macmillan, 1948.

Meyer, John R., *et al. The Economics of Competition in the Transportation Industry*. Cambridge, Mass.: Harvard University Press, 1959.

Mohring, Herbert D., and Mitchell Harwitz. *Highway Benefits: An Analytical Framework*. Evanston, Ill.: Northwestern University Press, 1962.

Shultz, Richard P. "Graphic Analysis of Waterway Capacity." (Conference preprint 409, American Society of Civil Engineers Transportation Engineering Conference), October 17–21, 1966.

Smith, Vernon L. *Investment and Production: A Study in the Theory of the Capital Using Enterprise*. Cambridge, Mass.: Harvard University Press, 1961.

Takacs, Lajos. *Introduction to the Theory of Queues*. New York: Oxford University Press, 1962.

U.S. Army. Corps of Engineers. *Regulations Prescribed by the Secretary of the Army for Ohio River, Mississippi River Above Cairo, Ill., and their Tributaries; Use, Administration, and Navigation* (as amended). Washington, D.C.: Government Printing Office, March 1968.

————. *Transportation Lines on the Mississippi River System and the Gulf Intracoastal Waterway.* Washington, D.C.: Government Printing Office. Annual.

————. *Waterborne Commerce of the United States.* Supplement to Part 5. *Domestic Inland Traffic Areas of Origin and Destination of Principal Commodities.* Washington, D.C.: Government Printing Office, 1956–63. Annual.

————. Ohio River Division. *Resistance of Barge Tows; Model and Prototype Investigations.* (Civil Works Investigations 814 and 835). Cincinnati, Ohio, August 1960.

————. ————. *Supplement to Resistance of Barge Tows; Model and Prototype Investigations.* (Report by L. A. Baier). Cincinnati, Ohio, June 1963.

U.S. Bureau of Mines. *Minerals Yearbook,* Vol. 2, *Fuels.* Washington, D.C.: Government Printing Office, 1956–63. Annual.

U.S. Congress. Senate. *Policies, Standards, and Procedures in the Formulation, Evaluation, and Review of Plans for the Use and Development of Water and Related Land Resources.* (Senate Document 97.) 87 Cong. 2 sess. Washington, D.C.: Government Printing Office, May 29, 1962.

————. ————. Select Committee on National Water Resources. *Water Resources Activities in the United States; Future Needs for Navigation.* 86 Cong. 2 sess. Washington, D.C.: Government Printing Office, 1960.

U.S. Department of Agriculture. *Agricultural Statistics.* 1956–63. Annual.

U.S. Department of Commerce. *The Annual Survey of Manufacturers.* 1956–63. Annual.

U.S. Interstate Commerce Commission. *State to State Distribution, Products of Mines, Traffic and Revenue, One Percent Sample of Terminations in the Years 1956–1963.* Washington, D.C.: Government Printing Office, 1956–63. Annual.

————. Bureau of Transport Economics and Statistics. *Transport Statistics in the United States, Part 5: Carriers by Water.* Washington, D.C.: Government Printing Office. Annual.